Niels Bohr:
Politics, Popularization and People

A booklet on the occasion of the World Year of Physics

Prepared by Finn Aaserud, General Editor of the Niels Bohr Collected Works

1st edition 2005
ISBN: 0444519149
Price: USD 19/ EUR 19

Coverphotograph: Niels Bohr, Albert Einstein, T. de Donder, O.W. Richardson, P. Langevin, P. Debye, A.F. Joffe and B. Cabrera, gathered to make plans for the 1933 Solvay Conference, Brussels, 1932
Photographer: Queen Elizabeth of Belgium

Preface

The World Year of Physics, which is celebrated in 2005, marks the hundredth anniversary of Albert Einstein's revolutionary publications in the 1905 volume of the German journal "Annalen der Physik". In addition to introducing the special theory of relativity, Einstein here made the radical proposal that the photoelectric effect might be explained by considering light as consisting not of waves, but of particles. The 1905 publications constituted a considerable triumph for the twenty-six year old physicist, who among colleagues and the general public alike came to be regarded as the greatest physicist of the twentieth century, closely followed by the Danish scientist Niels Bohr.

Six years younger than Einstein, Bohr had completed his gymnasium education in 1903 and had started his studies at the University of Copenhagen when Einstein's papers appeared. Bohr made his first fundamental contribution to physics in 1913, when he published his quantum theory of the atom in his so-called trilogy of papers in the British "Philosophical Magazine". When he received the Nobel Prize in physics for this work in 1922, Bohr went out of his way to emphasize that his contribution was part of a tradition in the new quantum physics beginning with Max Planck's introduction of the quantum concept in 1900 and continuing with Einstein's paper on the photoelectric effect as well as Ernest Rutherford's experimental discovery of the nuclear atom in 1911, which served as the empirical basis for Bohr's theory.

The year 2005 celebrates not only the hundredth anniversary of Einstein's breakthrough as a physicist, but also the eightieth anniversary of Werner Heisenberg's original formulation of quantum mechanics, one of the most important contributions to twentieth century physics. Bohr's renowned Institute for Theoretical Physics at the University of Copenhagen was central in the development leading to and arising from Heisenberg's theory, with Bohr himself teaching, advising and binding together a whole new generation of physicists. In particular, Bohr and Heisenberg, together with other younger physicists visiting the institute in Copenhagen, went on to develop what was to become known as the "Copenhagen Interpretation" of quantum mechanics, which still today is accepted by many, possibly most, physicists as the way to understand the new physics. Einstein's refusal to accept the statistical quantum mechanics as final provided an occasion to bring the two giants of physics together. Their famous discussions on the matter, with regard to which they were never reconciled, took place in particular at the 1927 and 1930 Solvay Conferences in Brussels. These encounters between Bohr and Einstein have been described in many places, notably Volume 7 of the Niels Bohr Collected Works, which contains Bohr's own detailed and fascinating account.

In the course of their careers Einstein and Bohr became public figures, to which they – again in different ways dictated by personality as well as circumstances – responded actively and conscientiously. In particular, both were genuinely concerned about the social and political responsibility of the physicist, especially in the wake of the atomic bomb, to the origins of which both had contributed in different ways.

After Niels Bohr's published contributions to physics and philosophy have been reproduced and contextualized in the first ten volumes of his Collected Works, the last two volumes of the series document his involvement, broadly speaking, in the political arena (Volume 11) as well as his endeavours towards science popularization and his extensive personal network – including Einstein – inside and outside the field of physics (Volume 12). These writings by Bohr serve to place the great Danish scientist and his contributions in perspective, and we are proud to be able to announce the publication of the last two volumes of the Niels Bohr Collected Works during the World Year of Physics.

Finn Aaserud
Niels Bohr Archive
December 2004

Contents

Preface III
Contents V
Earlier Volumes of the Niels Bohr Collected Works VI
Introduction 1
1. An "Open World" 2
2. Other Political and Social Involvements 3
3. Science and its Popularization 4
4. People 4

Volume 11, Part I: An "Open World" **7**
I Documents, World War II 9
1 Letter to John Anderson, 16 February 1941 10
2 Letter to Winston Churchill, 22 May 1944 14
3 Letter to Franklin D. Roosevelt, 7 September 1944 17
II *Science and Civilization,* The Times, 11 August 1945 21

Volume 11, Part II: Other Political and Social Involvements **27**
III *Science and its International Significance,* Danish Foreign Office Journal, No. 208 (May 1938) 61–63 29

Volume 12, Part I: Science and its Popularization **35**
IV *Atoms and Human Knowledge; Atomerne og vor Erkendelse,* Berlingske Tidende, 2 April 1949 37
Translation 43

Volume 12, Part II: People **53**
V *Professor Sir Ernest Rutherford and his Significance for the Recent Development of Physics; Professor Sir Ernest Rutherford og hans Betydning for Fysikens nyere Udvikling,* Politiken, 18 September 1920, pp. 7–8 55
Translation 60
VI *The Internationalist* [Albert Einstein], UNESCO Courier **2** (No. 2, March 1949) 1, 7 65
VII *Kirstine Meyer, b. Bjerrum: 12 October 1861 – 28 September 1941; Kirstine Meyer, f. Bjerrum: 12. Oktober 1861 – 28. September 1941,* Fysisk Tidsskrift **39** (1941) 113–115 71
Translation 75

World Year of Physics

EARLIER VOLUMES OF THE NIELS BOHR COLLECTED WORKS

In addition to the editor of each individual volume (see below), a General Editor has overseen the preparation of the entire series of the *Niels Bohr Collected Works.* The General Editors have been: Léon Rosenfeld (1904–1974), Volumes 1 to 3; Erik Rüdinger, Volumes 5 to 9 (Volume 7 jointly with Finn Aaserud); and Finn Aaserud, Volume 10. All volumes are published by North-Holland/Elsevier.

Vol. 1, *Early Work (1905–1911)* (ed. J. Rud Nielsen), 1972.

Vol. 2, *Work on Atomic Physics (1912–1917)* (ed. Ulrich Hoyer), 1981.

Vol. 3, *The Correspondence Principle (1918–1923)* (ed. J. Rud Nielsen), 1976.

Vol. 4, *The Periodic System (1920–1923)* (ed. J. Rud Nielsen), 1977.

Vol. 5, *The Emergence of Quantum Mechanics (Mainly 1924–1926)* (ed. Klaus Stolzenburg), 1984.

Vol. 6, *Foundations of Quantum Physics I (1926–1932)* (ed. Jørgen Kalckar), 1985.

Vol. 7, *Foundations of Quantum Physics II (1933–1958)* (ed. Jørgen Kalckar), 1996.

Vol. 8, *The Penetration of Charged Particles Through Matter (1912–1954)* (ed. Jens Thorsen), 1987.

Vol. 9, *Nuclear Physics (1929–1952)* (ed. Sir Rudolf Peierls), 1986.

Vol. 10, *Complementarity Beyond Physics (1928–1962)* (ed. David Favrholdt), 1999.

Introduction

by

Finn Aaserud

The pages following this Introduction contain selected writings from the last two volumes (Volumes 11 and 12) of the Niels Bohr Collected Works, which are currently being prepared for publication. The present Introduction is intended to place these writings in context – both generally and in relation to their appearance in the Collected Works.

The project to publish the Niels Bohr Collected Works was initiated under the leadership of Bohr's younger colleague and close collaborator, the Belgian-born physicist Léon Rosenfeld, the first volume appearing in 1972. Beginning with Volume 5, the general editorship of the series was taken over by Erik Rüdinger, Niels Bohr's last assistant. Rüdinger was in charge of the Niels Bohr Archive in 1985, when it was formally set up as an independent repository for Bohr's papers at the centennial of Bohr's birth. In the process, the institutional responsibility for the publication of the Collected Works was transferred from the Niels Bohr Institute to the Niels Bohr Archive. Therefore, the General Editor of the series changed once more when I succeeded Erik Rüdinger as director of the Niels Bohr Archive in 1989.

The preparation of each individual volume has been carried out by a special editor with expert knowledge of the relevant aspect of Bohr's work. So far ten volumes have been published, covering Bohr's contributions to physics (Volumes 1 to 9) and philosophy (Volume 10). Each volume consists of one or more numbered Parts in which the editor's Introduction to the topic at hand is followed by a main section consisting of Bohr's publications on this topic. While the Collected Works aim at completeness only with regard to Bohr's publications, selections of particularly relevant correspondence and manuscripts, as well as photographs, are also included.

Bohr's remaining publications will take up two more volumes, both of which are edited by the present writer. Volume 11, *The Political Arena*, starts by documenting Bohr's veritable crusade, begun during the Second World War, for what he termed an "open world" between nations. The volume goes on to deal with his various other political involvements, broadly speaking, as these are reflected in his publications. Volume 12, *Popularization and People*, includes a number of unique writings, many of which have not previously been published in English.

Each of Volumes 11 and 12 is divided into two Parts. The present booklet presents samples from all four of them, as described in the following sections.

1. AN "OPEN WORLD"

Almost as soon as he understood in 1943 that an atomic bomb would likely be built by the end of the Second World War, Bohr developed and promoted his view that the existence of the bomb, as he expressed it, would both facilitate and necessitate an "open world" between nations once the war was over. Bohr has described his ideas and efforts in this regard in his "Open Letter to the United Nations", published in June 1950. Although there are also other publications from Bohr's hand on this issue, especially from the period shortly after the Second World War, they are so few and far between as not to stand in any reasonable relation to the extent of Bohr's effort, which indeed constituted his main concern and occupation for the last twenty years of his life.

In addition to his publications proper, Bohr thus wrote a number of confidential memoranda to statesmen, laying out his ideas and recommending that they be put into practice. This was particularly the case during the Second World War, when Bohr's endeavours were subject to strict secrecy. Bohr continued to place hopes in his confidential approach even after the war, producing more confidential writings and discussing them with leading politicians. The "Open Letter" contains extensive quotations from those of Bohr's confidential writings that he considered most important. However, his total output of such writings is considerably larger and they were furthermore written and rewritten and rewritten again just as carefully as any of Bohr's publications proper. In this Part of the Collected Works it has therefore been deemed appropriate, as an exceptional case, to mix Bohr's published and unpublished writings.

The documents presented in the main section even include some particularly elaborate letters, into which Bohr put special work and thought. The examples of Bohr's writings chosen for reproduction below comprise three such letters. The first, addressed to the British Chancellor of the Exchequer and political leader of the British section of the atomic bomb project, Sir John Anderson, represents Bohr's first formulation in writing of how the existence of the atomic bomb provided an opportunity for the statesmen to improve international relations after the war was over. The other two letters, addressed respectively to Winston Churchill, the British Prime Minister, and Franklin D. Roosevelt, the President of the United States, were responses to the personal meetings Bohr was able to arrange with each of them. All of the letters show Bohr's strong commitment and how far he was able to bring his cause during the war. After the existence of the atomic bomb became public knowledge in August 1945, Bohr was allowed to present his views in print. He immediately published an article in The Times in London, which is also reproduced below. Here he only presented his general views, without mentioning how he had sought to carry them out during the foregoing two years. In this way, he could continue his confidential approaches to the statesmen even after the war.

Bohr's crusade to achieve an "open world" lasted until his death. In contrast to his scientific publications, his writings on this issue cannot be understood without relation to the general effort of which they were part. Therefore, Volume 11, Part I contains a substantial Appendix of notes and letters further illuminating Bohr's efforts. As another exception in relation to other volumes of the Collected Works, comments between third parties about Bohr's efforts have been included in addition to letters to and from Bohr himself. In order to bring the whole story together, the Introduction is written with an attention to detail that is unusual for the Collected Works. The author is grateful to Aage Bohr, who assisted his father in his political activities and followed them at close hand, for invaluable help in preparing this Part of the Collected Works, and particularly for his help with all aspects of writing the Introduction.

2. OTHER POLITICAL AND SOCIAL INVOLVEMENTS

Throughout his life Bohr involved himself in several political and social activities, in addition to those described in Part I. The purpose of the Niels Bohr Collected Works is to document these activities only in so far as they produced genuine publications. Therefore, some major activities which did not lead to published documents will not be represented. These include, among others, Bohr's involvement in the 1920s in the expansion and relocation of the Danish National Museum and his role in the 1950s in the establishment of the European nuclear research facility of CERN. Yet, several of Bohr's political and social activities did involve publications, and it is these that will constitute Part II of Volume 11 of the Collected Works. These activities range from Bohr's presidency of the Royal Danish Academy of Arts and Sciences between 1939 and 1962 to addresses on various occasions.

The content of several of these publications reflects the political or social context only to a small degree and sometimes not at all. In many instances, Bohr used the opportunity provided by a public address to present some question in physics to non-physicists or to describe the philosophical implications of modern physics. In such cases, the occasion, not the content, of the article has determined its placement in this Part of the Collected Works. The Introduction to Volume 11, Part II will serve to document the context for Bohr's contributions when his publications fail to do so. The article chosen for this booklet arose in connection with the twenty-fifth anniversary of Bohr's presentation of his atomic model in 1913, which coincided with a substantial physical expansion of his institute as well as the awarding to Bohr of the Danish Medal of Merit in Gold. The celebration was attended by representatives of the Danish cultural and political elite and received substantial attention in the Danish Press. While the talk (in Danish) that Bohr gave to the invited audience at his institute was not published, a different address (in English) prepared for international radio broadcasting did appear in print. Here Bohr characteristically did not pay particular attention to his institute's past accomplishments or future plans, but presented instead the recent development of physics in general terms. Nevertheless, the occasion constituted in effect not only a celebration of his discovery twenty-five years before, but also of the nearly two decades old Institute for Theoretical Physics at the University of Copenhagen, the establishment and leadership of which must be considered one of Bohr's major accomplishments outside his scientific and intellectual contributions.

3. SCIENCE AND ITS POPULARIZATION

The main content of Volume 12, Part I comprises those of Bohr's popular accounts of modern physics and its epistemological implications that belong neither in Volume 11 nor in earlier volumes (notably Volumes 7 and 10). Bohr's radio address from 1949 to Danish and Norwegian gymnasium students constitutes a particularly instructive example of his concerted mission to make the epistemological lesson of quantum mechanics understandable to the general public. It is reproduced below, both in Danish facsimile and English translation.

4. PEOPLE

In the course of his career, Bohr published articles about several of his predecessors in science, about contemporary physicists as well as about people outside the field of physics. Without in any way exhausting Bohr's considerable social and scientific network, taken together these articles give a better glimpse of it than any other set of his published writings.

Some of Bohr's tributes and obituaries have been published in earlier volumes of his Collected Works, where, as is also the case for many of his popular writings, they were included to illuminate the various topics under discussion there. The remaining ones, numbering more than fifty, are reproduced in Volume 12, Part II of the Collected Works.

In addition to Bohr's writings about other people, this Part of the Collected

Works includes autobiographical pieces in the form of contributions written on request in connection with honorary distinctions as well as interviews in the media. These writings provide a rare autobiographical element which is invaluable for understanding Bohr and the background for his several publications honouring others.

Bohr's most substantial tributes were written to his revered mentor Ernest Rutherford and to his main intellectual sparring partner Albert Einstein. These contributions have been reproduced in Volumes 10 and 7, respectively. However, Bohr's writings on Rutherford and Einstein are not limited to these two articles. His earliest publication on Rutherford was written for a major Danish newspaper when Rutherford visited the Danish capital for the first time in 1920. Three decades later Bohr gave a radio talk on Einstein in celebration of the latter's seventieth birthday. These two contributions have been chosen for this booklet to represent Bohr's published writings in honour of his colleagues.

Bohr's ties to the Danish society in which he grew up were strong and, with the exception of his forced exile during the last two years of the Second World War, he lived and worked in Denmark throughout his life. Several of his tributes and obituaries were written in honour of members of his Danish circle. The contribution reproduced in this booklet is an article devoted to Kirstine Meyer, one of the first two women (the other was Hanna Adler, Bohr's maternal aunt) to obtain a Master's degree in physics at the University of Copenhagen in 1892. Meyer became a major force in Danish secondary physics education as well as a historian of physics. She was one of Bohr's strong supporters in his quest for a professorship at the University of Copenhagen, and their relationship did not diminish when Bohr was no longer in need of such help. Bohr's article was originally intended to celebrate Meyer's eightieth birthday, but became in effect an obituary when she died shortly before that occasion. Like several other writings in honour of fellow Danes, it illustrates Bohr's closeness to his home milieu.

VOLUME 11, PART I

AN "OPEN WORLD"

Niels Bohr (left) greets John Anderson on his visit to Copenhagen in 1947. Centre, Alec Randall, the British ambassador to Denmark.

Photographer: H. Lund Hansen.

I. DOCUMENTS, WORLD WAR II

Document 1:
LETTER TO ANDERSON, 16 February 1944
[Typewritten]

Washington, D.C. February 16th 1944.

PERSONAL. MOST SECRET.

The Rt. Hon. Sir John Anderson M.P.
Chancellor of the Exchequer.
Westminster, London.

Dear Sir John.

Through your introduction I was, as you will know, received by Lord Halifax shortly after my first arrival in Washington. On my return from a three weeks visit to "Y", I called again at the Embassy and have in the last weeks had various conversations with the Ambassador and with Sir Ronald Campbell who, according to the Ambassador's request was brought into the discussion of the matter. I understand that Lord Halifax will himself communicate with you, but I presume that you may be interested in a brief report about my experiences and endeavours since my departure from London.

As regards my own status no change has so far been proposed from the American side, and my stay in "Y" took place according to an invitation from General Groves who, without any special commitments except for security purposes, has agreed that I and my son go out to "Y" on temporary visits for consultations with the American and British scientists working there.

I was most impressed by the magnitude of the effort and the progress of the work already achieved. Thus all problems concerning the mechanism of the fundamental processes have been thoroughly investigated and in particular it has been proved that in accordance with the theoretical expectations no disturbing time delay in the neutron emission occurs. As regards the scientific basis no doubt is therefore left as to the realizability of the project.

Also on the preparations for the construction of the actual weapon a great amount of work has been done in "Y" and various ingenious methods by which smaller amounts of material than previously estimated may effect a release of nuclear energy on a large scale have recently been proposed. Much experimental work, however, will still be needed to test the efficiency of such methods.

In view of the decisive importance that all such preparations are completed before the material is at hand I have on my return to Washington in conversations with General Groves stressed the urgency that just this aspect of the work be promoted to the outmost, and he agrees upon the necessity of extending the work at "Y" with new facilities for carrying out large scale experiments on the latest devices for the assembly of the weapon. I know that Dr. Chadwick has proposed that the British collaboration on the project, not least in "Y", if possible be still further extended, and I wish most heartily to support such advice.

In Washington I have been in close connection with British scientists who frequently come for meetings from the various places where the work on the production of the materials take place and to which I have hitherto had no access. Through discussions, especially with Dr. Oliphant who is here on his way to England, I have obtained most encouraging information about the recent great progress of these truly gigantic efforts, which has secured an impending realization of the whole project.

This situation certainly calls for the most thorough attention to all implications of the project, and ever since our last conversation in London during which you showed me the confidence of indicating to me your concern about this aspect of the matter, such problems have continually been on my mind.

I know that the question of control has already been considered within your committee, but the more I have learned and thought about the possible development in this new field of science and technique, the more I am convinced that no kind of customary measures will suffice for this purpose and that no real safety can be achieved without a universal agreement based on mutual confidence.

An effective control would in fact not only involve intricate technical and administrative problems, but would also demand such concessions regarding exchange of information, and openness about industrial efforts including military preparations, as would hardly be conceivable unless at the same time all partners were assured of a compensating guarantee of common security against dangers of unprecedented acuteness.

In this connection it might even be necessary temporarily to renounce the great promises of the project as regards peaceful industrial development and leave the exploitation of such possibilities until by universal consent it is deemed that an established control will prevent any illegitimate use of the powerful new materials.

All such details will depend on technical developments which cannot be surveyed at the moment, but the main point of the argument is that the impending realization of the project would not only seem to necessitate, but should also, due to the urgency of confidence, facilitate, a new approach to the problem of international relationship.

I am well aware that I here enter upon questions about which only the responsible statesmen who alone have insight in the political possibilities are able to judge, but stressing the uniqueness of the situation which the progress of science has brought about I venture to point to the favourable consequences which might result from an early initiative from the side which by good fortune has obtained a lead in the efforts to master forces of nature hitherto beyond human reach.

Such an initiative, aimed at preventing a future competition about the formidable weapon should in no way impede the importance of the project for immediate military objectives, but might by strengthening the confidence within the United Nations turn this unparalleled enterprise to lasting benefit for the common cause.

Of course, I appreciate the necessity of separating sharply between the actual work on the project and its possible implications of general political character, just as I realize the serious complications for the handling of the whole situation which might arise from discussions of such implicatons even within the circle of officials and technicians charged with the development of the project.

I have, therefore, been most grateful for the privilege you have given me of bringing my impressions to the notice of Lord Halifax who, as well as Sir Ronald Campbell, has taken great interest in the matter. I understand that the Ambassador shares the view that the situation requires an early attention in order that no opportunity be lost.

Yours sincerely

N. Bohr

President Franklin D. Roosevelt and Prime Minister Winston Churchill, c. 1944.

Document 2:
LETTER TO CHURCHILL, 22 May 1944
[Typewritten]

22nd May 1944.

The Rt. Hon. Winston S. Churchill, C.H., M.P.

Sir,

In accordance with your kind permission, I have the honour to send you a brief report about my impressions of the great Anglo–American enterprise, in the scientific aspects of which I have been given the opportunity to participate together with my British colleagues.

The principles on which the enormous energy stored in the nuclei of atoms may be released for practical purposes were, as a result of international scientific collaboration, already perceived in outline before the war and are, therefore, common knowledge to physicists all over the world. It was, however, by no means certain whether the task would surpass human resources, and it was therefore a revelation to me, on my arrival in England last October, to learn with what courage and foresight the effort had been undertaken and what an advanced stage the work had already reached.

In fact, what until a few years ago might be considered as a fantastic dream is at present being realized within great laboratories and huge production plants secretly erected in some of the most solitary regions of the United States. There a larger group of physicists than ever before collected for a single purpose, working hand in hand with a whole army of engineers and technicians, are preparing new materials capable of an immense energy release, and are developing ingenious devices for the most effective use of these materials.

To everyone who is given the opportunity to see for himself the refined laboratory equipment and the gigantic production machinery, it is an unforgettable experience, of which words can only give a poor impression.

Moreover it was to me a special pleasure to witness the most harmonious and enthusiastic co-operation between the British and American colleagues, and on my departure I was expressly asked by the leaders of the American organization to convey their genuine appreciation of the help they are receiving, on an ever increasing scale, from their British collaborators.

I will not tire you with any technical details, but one cannot help comparing the situation with that of the alchemists of former days, groping in the dark in their vain efforts to make gold. To-day physicists and engineers are, on the basis of firmly established knowledge, controlling and directing violent reactions by which new materials far more precious than gold are built up, atom by atom.

These processes are in fact similar to those which took place in the early stages of development of the universe and still go on in the turbulent and flaming interior of the stars.

The whole undertaking constitutes, indeed, a far deeper interference with the natural course of events than anything ever before attempted, and it must be realized that the success of the endeavours has created a quite new situation as regards human resources.

The revolution in industrial development which may result in coming years cannot at present be surveyed, but the fact of immediate preponderance is, that a weapon of devastating power far beyond any previous possibilities and imagination will soon become available.

The lead in the efforts to master such mighty forces of nature, hitherto beyond human reach, which by good fortune has been achieved by the two great free nations, entails the greatest promises for the future. The responsiblity for handling the situation rests, of course, with the statesmen alone. The scientists who are brought into confidence can only offer the statesmen all such information about technical matters as may be of importance for their decisions.

In this connection it is significant that the enterprise, immense as it is, has still proved to demand a much smaller effort than might have been anticipated, and that the development of the work has continually revealed unsuspected possibilities for facilitating the production of the materials and for intensifying their effects.

These circumstances obviously have an important bearing on the question of an eventual competition about the formidable weapon, and on the problem of establishing an effective control, and might therefore perhaps influence the judgment of the statesmen as to how the present favourable situation can best be turned to lasting advantage for the cause of freedom and world security.

I hope you will permit me to say that I am afraid that, at the personal interview with which you honoured me, I may not have given you the right impression of the confidential conversation in Washington on which I reported. It was, indeed, far from my mind to venture any comment about the way in which the great joint enterprise has been so happily arranged by the statesmen; I wished rather to give expression to the profound conviction I have met everywhere on my journey that the hope for the future lies above all in the most brotherly friendship between the British Commonwealth and the United States.

It was just in this spirit of co-operation that the President's friend believing the matter to be of the highest importance for the two countries, and knowing that, at the Chancellor's request, I was coming to England for technical consultations, entrusted me, in strictest confidence, to convey to you, that the President is deeply occupied in his own mind with the stupendous consequences of the project, in which he sees grave dangers, but also unique opportunities, and that he hopes together with you to find ways of handling the situation to the greatest benefit of all mankind.

Most respectfully,
[Niels Bohr]

Document 3:
LETTER TO ROOSEVELT, 7 September 1944
[Carbon copy]

[Editor's comment: A different version of the letter, deposited in the Frankfurter file of the Oppenheimer Papers, Library of Congress, Washington, D.C., seems to be an earlier draft.]

TOP SECRET

Washington, September 7th 1944.

My dear Mr. President.

I wish to thank you most heartily for the honour and confidence you showed me by receiving me and talking with me of your concern about the great and urgent problems raised by the recent extraordinary development of physical science.

The impending realization of the prospects of releasing atomic energy on a large scale, secured by the great pioneer work in this country, is surely destined deeply to influence the future of mankind. But, as you appreciate so fully, the bright promises which the wonderful adventure entails may be overshadowed by most ominous threats to human security unless in due time international agreement can be obtained as regards an effective control of the new formidable weapon.

As a physicist who has been given the privilege to follow the latest development at close hand, I was, however, most grateful for the opportunity to bring before you some considerations concerning the technical aspects of the unparallelled enterprise with special regard to the problem of control measures and the possibilities of competing efforts in other countries.

In the last respect it should be borne in mind that, as a result of a fruitful international scientific co-operation, the principles on which a large scale release of atomic energy can be effected were, at any rate in outlines, perceived already before the war and it is known that attempts of realizing the prospects have since been made in various countries.

It seems certain that the American and British effort undertaken with so happy a foresight and backed with such great resources has achieved a decisive lead. Still, it must be realized that any information, however scanty, about the success of the work in this country which may have leaked out is likely to have caused the utmost intensification of similar efforts elsewhere.

Fortunately the course of the war has probably removed any danger of a military use of the purposeful endeavours on such lines made in Germany. One must, however, be prepared that with the final defeat of Germany all expert knowledge and technical experience collected there will presumably become available in equal measure for the great victorious nations.

It is from the point of view of proper timing that I venture to suggest that the technical factors of the situation make the present moment most favourable for considerations of the question of control by the friendly governments most concerned, lest opportunities be forfeited of forestalling a fateful competition about the new weapon and of turning the great triumph of science and engineering to lasting benefit for the common cause.

Of course, it is wholly outside my province to suggest the procedure appropriate for such delicate problems of statesmanship. But as a scientist it occurs to me that in this unique situation pre-war scientific connections may prove helpful in conveying, with entire regard to security, an understanding of how much would be at stake should the great prospects of atomic physics materialize, and in preparing an adequate realization of the great benefit which would ensue from a whole-hearted co-operation on effective control measures.

I wish once more to thank you for the great kindness you showed me, and for the wish you expressed to see me again at some later occasion. I need not assure you that I shall be most happy to respond to the honour of a call from you.

Most respectfully yours
[Niels Bohr]

The President.

Niels Bohr Collected Works, Vols. 11 & 12

II. SCIENCE AND CIVILIZATION

The Times, 11 August 1945

SCIENCE AND CIVILIZATION (1945)

Versions published in English and Danish

English: Science and Civilization

A *Energy from the Atom – An Opportunity and a Challenge – The Scientist's View,* The Times, 11 August 1945

B "One World or None" (eds. D. Masters & K. Way), McGraw-Hill, New York 1946, pp. IX–X

Danish: Videnskab og Civilisation

C *Energi fra Atombomber – En Mulighed og en Udfordring – Videnskabsmændenes Syn,* Berlingske Aftenavis, 11 August 1945, pp. 1–2

D "Een Verden eller ingen" (eds. D. Masters & K. Way), Haase, Copenhagen 1946, pp. 11–15

All of these versions agree with each other, with the exception that *C* and *D* are different translations.

REPRINTED FROM

THE TIMES

Saturday August 11 1945

SCIENCE AND CIVILIZATION

By Niels Bohr

The possibility of releasing vast amounts of energy through atomic disintegration, which means a veritable revolution of human resources, cannot but raise in the mind of everyone the question whither the advance of physical science is leading civilization. While the increasing mastery of the forces of nature has contributed so prolifically to human welfare, and holds out even greater promises, it is evident that the formidable power of destruction which has come within reach of man may become a mortal menace unless human society can adjust itself to the exigencies of the situation. Civilization is presented with a challenge more serious, perhaps, than ever before, and the fate of humanity will depend on its ability to unite in averting common dangers and jointly to reap the benefit from the immense opportunities which the progress of science offers.

In its origin science is inseparable from the collecting and ordering of experiences, gained in the struggle for existence, which enabled our ancestors to raise mankind to its present position among the other living beings which inhabit our earth. Even in highly organized communities where, within the distribution of labour, scientific study has become an occupation by itself, the progress of science and the advance of civilization have remained most intimately interwoven. Of course, practical needs are still an impetus to scientific research, but it need hardly be stressed how often technical developments of the greatest importance for civilization have originated from studies aimed only at augmenting our knowledge and deepening our understanding. Such endeavours know no national borders and where one scientist has left the trail another has taken it up, often in a distant part of the world. For long scientists have considered themselves as a brotherhood working in the service of common human ideals.

In no domain of science have these lessons received stronger emphasis than in the exploration of the atom, which just now is bearing consequences of such overwhelming practical implications. As is well known, the roots of the idea of atoms as the ultimate constituents of matter go back to ancient thinkers searching for a foundation to explain the regularity which, in spite of all variability, is ever more clearly revealed by the study of natural phenomena. After the Renaissance, when science entered so fertile a period, atomic theory gradually became of the greatest importance for the physical and chemical sciences, although, until half a century ago, it was generally accepted that, due to the coarseness of our senses, any direct proof of the existence of atoms would always remain beyond human scope. Aided, however, by the refined tools of modern technique, the development of the art of experimentation has removed such limitation and even yielded detailed information about the interior structure of atoms.

In particular, the discovery that almost the entire mass of the atom is concentrated in a central nucleus proved to have the most far-reaching consequences. Not only did it become evident that the remarkable stability of the chemical elements is due to the immutability of the atomic nucleus when exposed to ordinary physical agencies but a novel field of research was opened up by the study of the special conditions under which disintegrations of the nuclei themselves may be brought about. Such processes, whereby the very elements are transformed, were found to differ fundamentally in character and violence from chemical reactions, and their investigation led to a rapid succession of important discoveries through which ultimately the possibility of a large-scale release of atomic energy came into sight. This progress was achieved in the course of a few decades, and was due not least to most effective international cooperation. The world community of physicists was, so to say, welded into one team, rendering it more difficult than ever to disentangle the contributions of individual workers.

The grim realities which are being revealed to the world in these days will no doubt, in the minds of many, revive terrifying prospects forecast in fiction. With all admiration for such imagination, it is, however, most essential to appreciate the contrast between these fantasies and the actual situation confronting us. Far from offering any easy means to bring destruction forth, as it were by witchcraft, scientific insight has on the contrary made it evident that use of nuclear disintegration for devastating explosions demands most elaborate preparations, involving a profound change in the atomic composition of the materials found on earth. The astounding achievement of producing an enormous display of power on the basis of experience gained by the study of minute effects, perceptible only by the most delicate instruments, has, in fact, besides a most intensive research effort, required an immense engineering enterprise, strikingly illuminating the potentialities of modern industrial development.

Indeed, not only have we left the time far behind where each man, for self-protection, could pick up the nearest stone, but we have even reached the stage where the degree of security offered to the citizens of a nation by collective defence measures is entirely insufficient. Against the new destructive powers no defence may be possible, and the issue centres on world-wide cooperation to prevent any use of the new sources of energy which does not serve mankind as a whole. The possibility of international regulation for this purpose should be ensured by the very magnitude and the peculiar character of the efforts which will be indispensable for the production of the new formidable weapon. It is obvious, however, that no control can be effective without free access to full scientific information and the granting of the opportunity of international supervision of all undertakings which, unless regulated, might become a source of disaster.

Such measures will, of course, demand the abolition of barriers hitherto considered necessary to safeguard national interests but now standing in the way of common security against unprecedented dangers. Certainly the handling of the precarious situation will demand the good will of all nations, but it must be recognized that we are dealing with what is potentially a deadly challenge to civilization itself. A better background for meeting such a situation could hardly be imagined than the earnest desire to seek a firm foundation for world security, so unanimously expressed from the side of all those nations which only through united efforts have been able to defend elementary human rights. The extent of the contribution which an agreement about this vital matter would make to the removal of obstacles to mutual confidence and to the promotion of a harmonious relationship between nations can hardly be exaggerated.

In the great task lying ahead, which places upon our generation the gravest responsibility towards posterity, scientists all over the world may offer most valuable services. Not only do the bonds created through scientific intercourse form some of the firmest ties between individuals from different nations, but the whole scientific community will surely join in a vigorous effort to induce in wider circles an adequate appreciation of what is at stake and to appeal to humanity at large to heed the warning which has been sounded. It need not be added that every scientist who has taken part in laying the foundation for the new development, or has been called upon to participate in work which might have proved decisive in the struggle to preserve a state of civilization where human culture can freely develop, is prepared to assist in any way open to him in bringing about an outcome of the present crisis of humanity worthy of the ideals for which science through the ages has stood.

Printed by The Times Publishing Company, Limited, Printing House Square, London, E.C.4, England.

VOLUME 11, PART II

OTHER POLITICAL AND SOCIAL INVOLVEMENTS

III. SCIENCE AND ITS INTERNATIONAL SIGNIFICANCE

Danish Foreign Office Journal, No. 208 (May 1938) 61–63

The annual staff picture taken in connection with Bohr's birthday, 7 October, Copenhagen, 1938.

DANISH FOREIGN OFFICE JOURNAL

COMMERCIAL
& GENERAL REVIEW

NUMBER 208 · MAY 1938

SCIENCE *and its* INTERNATIONAL SIGNIFICANCE

Broadcast lecture by
Professor NIELS BOHR, Copenhagen

The announcer: *This is Copenhagen, Denmark, calling.*

This programme is being sent from the Theoretical Physics Institute of the University. It is dedicated to the head of the institute, Professor Niels Bohr, the famous Danish physicist and Nobel Prize winner. An American guest in our institute, Professor Louis Turner of Princeton University, has consented to give a few introductory words. Will you please take the microphone, *Professor Turner?* . . . Yes, with the greatest pleasure:

Twenty-five years ago, on the fifth of April, nineteen hundred and thirteen, Professor Niels Bohr finished his first paper on the Bohr theory of the structure of the atom. On the basis of the late Lord Rutherford's discovery of the nucleus of the atom, and Max Planck's discovery that radiation is emitted in separate quanta, Niels Bohr created his fundamental theory of the atom. For the past twenty five years this theory has been the guide in all progress in atomic physics.

The institute of Theoretical Physics was built for Professor Bohr in nineteen hundred and twenty in order to bring about an intimate collaboration between theoretical and experimental physics. To-day a new department of this institute has been officially opened in the presence of representatives of the Danish government and other notables.

In the last few years the greatest progress in physics has been achieved in the study of the structure of the atomic nuclei and of their disintegrations. Professor Bohr's recent contributions in this field are also of fundamental significance. This newly opened part of the laboratory contains modern equipment for experimental studies of this sort. The buildings and their equipment are gifts of the Rockefeller Foundation and of various Danish foundations. The more than one hundred colleagues from America and other foreign lands who have had the happy experience of working here with Professor Bohr, will envy me my good fortune in being here for this celebration. I am sure that I speak for them as well as for myself in offering congratulations and best wishes for another quarter century of splendid work. Professor Bohr will now speak to you.

Professor Bohr:

I am very thankful for the great honour shown me by the invitation to say a few words about the exploration of the wonderful world of atoms to listeners in the large English speaking communities on both sides of the Atlantic. It especially offers me a most welcome opportunity to give expression for the indebtedness all scientists in this field owe to the pioneer work of men from these great nations which have contributed so much to human civilisation. Nothing can indeed remind us more strongly of the unity of mankind than the evolution of science which, from primitive endeavours to collect and order experience, has grown to be a most powerful source of progress for humanity.

To this great work all civilized nations have given their share, and, approaching our subject, we need only think of men like *Franklin, Volta, Ørsted, Ampère, Faraday* and *Hertz,* to realize how international cooperation has created the basis for the wonders of electrical engineering, which have reshaped in an undreamt measure our daily life and at the same time given us new means to unveil the secrets of nature. The names of *Röntgen* and *Curie* will here suffice to recall, how even most beneficial aids to the relief of human suffering may be the immediate result of pure scientific inquiry into the physical nature of our world.

Of course, in a short speech it would be quite impossible to give any detailed account of the marvellous development of atomic physics in our days, and I shall therefore only recall a few points especially suited to illustrate the decisive role cooperation here has played. As everyone knows, a corner- stone in this development was the discovery of the electron, which was traced as a constituent element of all matter by the pioneer-work of *Sir J. J. Thomson* and his school in the old University of Cambridge, and the charge and mass of which was finally measured with such astounding accuracy through *Millikan*'s ingenious experiments in Chicago. It is also of common knowledge, that *Lord Rutherford*'s fundamental research on radioactivity led him to the discovery of the atomic nucleus which in so unsuspected manner completed our knowledge of the constitution of the atom. This discovery, however, revealed at the same time the impossibility of explaining many characteristic properties of atoms by means of the ordinary laws of physics, which apply to the behaviour of matter in bulk.

Still, due of the fundamental discovery, a few years earlier, of the elementary quantum of action, to which *Planck* was led through his penetrating analysis of black body radiation, physicists were in no way unprepared for such a situation. Planck's discovery meant indeed the disclosure of an essential new feature of atomicity in the laws of nature, quite foreign to the point of view of classical physics and supplementing most unsuspectedly the old doctrine of the restricted divisibility of matter. Thus was initiated a new epoch in natural philosophy, and it needs hardly be recalled that *Einstein,* in the very same year in which he published his first paper on relativity, also contributed most fruitfully to the foundations of atomic physics by pointing out, how the existence of the quantum of action offered an immediate explanation of a number of phenomena which hitherto had frustrated the efforts of scientists.

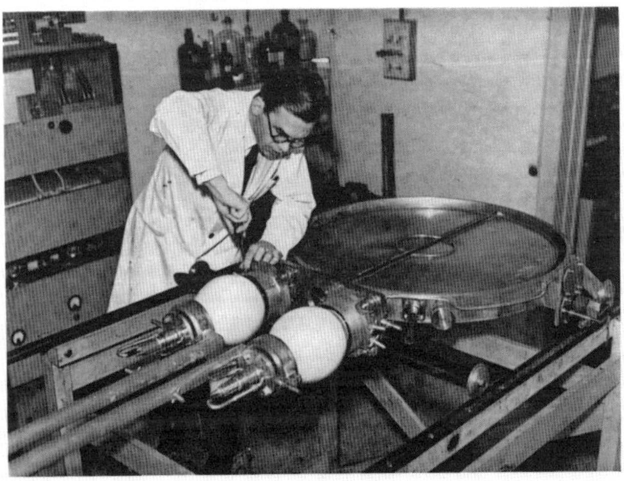

Electrodes for the Copenhagen Cyclotron granted to Professor Bohr by the Thomas B. Thrige Fund, Denmark

One had therefore not to look far for a clue to the solution of the new dilemma with which Rutherford's discovery confronted us, and it is a special pleasure for me here to stress that if, twentyfive years ago, I had the good fortune to give a modest contribution to this development, it was above all thanks to the hospitality I then as a young man, enthusiatic for the great promises of the new theoretical outlooks, enjoyed in the famous laboratories of England, where the fundamental experimental discoveries were just being made. In particular I think with grateful emotion of the unique friendliness and straightforwardness with which *Rutherford,* in the mindst of his unceasing creative activity, was always prepared to listen to any student, behind whose youthful inexperience he perceived a serious interest.

To develope on the foundation laid by *Planck* and *Rutherford* a rational theory of atomic constitution claimed, however, a clarification of the principles of atomic mechanics, which was gradually accomplished by a number of experimental as well as theoretical investigations, in which physicists from countries all over the world have taken effective part. The astounding rapidity of this brilliant development has in fact only been made possible by a most intensive international cooperation that has been facilitated not least by the liberality with which *the Rockefeller Foundation* has permitted thousands of promising young men to complete their education and enlarge their outlook by temporary visits to institutes abroad, where they at the same time have brought new impulses. Also in this Institute we have had the pleasure and benefit of the visit of a large number of Rockefeller fellows, and it is perhaps not without interest to mention that the first among them and, I think, actually the very first regular Rockefeller fellow, was *Heisenberg,* whose fundamental contributions to atomic mechanics are so well-known. Likewise it is a special pleasure for us to think of the stay here, as a fellow of *the American-Scandinavian Foundation,* of Professor *Urey,* whose later discovery of the heavy hydrogen has opened such important new fields of research.

In recent years we have entered a new era in physical and chemical science initiated by another of the great discoveries with which *Rutherford's* genius presented us. Thus, in nineteen hundred and nineteen *Rutherford* succeeded in proving that the nucleus of the atom, the intangibility of which under all ordinary conditions is the very basis for the constancy of the elements, may under the extreme circumstances afforded by the bombardement with high-speed particles, itself undergo disintegrations. This modern revival of the old dreams of the alchemists has, in the last few years, led to a succession of most startling discoveries regarding the ultimate constituents of matter and has already found very important practical applications, especially through the discovery of artificial radioactivity by the daughter and son-in-law of the immortal Madame Curie. Above all this discovery has given us entirely new means of studying the incessant exchange of matter within living organisms by means of the indicator method originated by Professor *Hevesy,* the privilege of whose collaboration we enjoy in this Institute.

Part of accelerating tube for nuclear disintegration experiments.

The intense world-wide interest in this young nuclear physics has given rise to great efforts in developing ever more powerful tools for this kind of research, ranging from the construction on a gigantic scale of electrostatic machines akin to those of Franklin's days to the design of electromagnetic accelerators of atomic particles on quite new principles. On my recent visit to America I received an unforgetful impression of the success of these courageous efforts, and it is a most pleasant duty on this occasion to express my special indebtedness towards the ingenious inventor of the cyclotron, Professor *Lawrence* of Berkeley, who has so generously put his precious experience at the disposal of this Institute as well as of many other laboratories all over the world. In building such complicated and costly apparatus, the mutual aid of scientific investigation and practical industrial activity is felt particularly strongly on account of the technical perfection required as well as of the necessary financial support. In our case this support has been most liberally granted by the *Rockefeller Foundation* and by Danish foundations which likewise owe their sources to industries which again depend essentially on science. In concluding this address I wish to give expression for the hope, that the example of the fruitful cooperation in science, independent of national borders and prejudices, may stimulate efforts to contribute by peaceful collaboration also in other fields of human culture to the common progress of mankind.

VOLUME 12, PART I

SCIENCE AND ITS POPULARIZATION

Niels and Margrethe Bohr with their grandchildren on Bohr's seventieth birthday, Copenhagen, 1955. Photographer: Holt and Madsen.

IV. ATOMS AND HUMAN KNOWLEDGE

ATOMERNE OG VOR ERKENDELSE
Berlingske Tidende, 2 April 1949

Talk on Danish radio for
Danish and Norwegian gymnasium students, 1 April 1949

TEXT AND TRANSLATION

ATOMS AND HUMAN KNOWLEDGE (1949)

Versions published in Danish

Atomerne og vor erkendelse
A Berlingske Tidende, 2 April 1949
B Vor Viden **33** (1950) 123

B is an abbreviated and slightly edited version of *A*

Niels Bohr:

Atomerne og vor erkendelse

Efter anmodning har professor Bohr overladt os manuskriptet til det foredrag, han i gaar har holdt i skoleradioen for danske og norske gymnasieelever

DET er en stor glæde for mig at faa en saadan lejlighed til at komme i forbindelse med den ungdom i Danmark og Norge, der gennem uddannelse i gymnasierne forbereder sig paa en gerning i samfundets tjeneste. Titlen paa mit foredrag „atomerne og vor erkendelse" lyder maaske meget lærd, men jeg skal forsøge at fortælle lidt om det indblik i atomernes verden, som er vundet ved en af de mest æventyrlige rejser af menneskeaanden i ukendte egne, hvor vi hele tiden har mødt store overraskelser og maattet overvinde mange vanskeligheder. De erfaringer, vi paa denne rejse har samlet, betyder et stort fremskridt i vor beherskelse af naturen, men samtidig har vi vundet en almen menneskelig belæring, som har bud til alle, baade ældre og yngre.

Tanken om, at alle legemer, som vi ser eller føler paa, er opbygget af atomer, som ved deres lidenhed unddrager sig direkte iagttagelse, gaar jo langt tilbage, og allerede græske filosoffer som Demokrit forstod, at det for at gøre rede for elementernes bestandighed var nødvendigt at antage en grænse for stoffernes delelighed. Naar f. eks. almindeligt vand ved fordampning kan overføres i en helt anden tilstandsform, og dampen derefter igen kan fortættes til vand med ganske samme egenskaber som før, ville dette jo ikke være til at begribe, hvis vi havde at gøre med fuldstændig forvandling af stofferne, men forholdet kan forklares helt simpelt, naar vi antager, at det, der sker ved fordampningen, er blot, at de tæt sammenpakkede smaadele i vandet fjernes fra hinanden, saa de for en tid kan bevæge sig mere frit imellem hverandre, og at der ved fortætningen ikke sker andet, end at molekylerne igen pakkes sammen, og deres bevægelighed begrænses.

Saadanne forestillinger viste sig efterhaanden yderst frugtbare og oplysende paa mange forskellige omraader, og især vandt vi i begyndelsen af forrige aarhundrede gennem Daltons værk en simpel forklaring af de konstante mængdeforhold, hvori grundstofferne indgaar i kemiske forbindelser. I et grundstof antager vi jo, at alle atomerne er ens, medens vi i en kemisk forbindelse har smaadele eller molekyler, der er sammensat af atomer af forskellige grundstoffer. F. eks. er vand en forbindelse af grundstofferne brint og ilt, som vi siger paa dansk, eller vandstof og surstof paa norsk, eller med betegnelser, som bruges saa mange andre steder, og som derfor ogsaa er i færd med at blive indført hos os: hydrogen og oxygen. Et vandmolekyle bestaar af to atomer hydrogen og eet atom oxygen, men vi har ogsaa en anden forbindelse af disse grundstoffer, brintoverilte, eller hydrogenperoxyd, hvis molekyler bestaar af to hydrogenatomer og to oxygen-atomer, og ved dannelse af dette stof vil derfor for samme mængde hydrogen indgaa netop den dobbelte mængde oxygen som ved dannelsen af vandet.

Omend paa mange saadanne maader atomteorien viste sig mere og mere uundværlig og overbevisende, var det dog lige op til vort aarhundrede en almindelig anskuelse, at det aldrig ville være muligt at føre direkte bevis for atomernes eksistens, idet man gik ud fra, at vore sansers grovhed ville udelukke enhver paavisning af tilstedeværelsen og virkningen af de enkelte atomer. Under fysikkens udvikling har vi imidlertid faaet mere og mere forfinede instrumenter til hjælp for vore sanser, og f. eks. har vi allerede forlængst ved mikroskopets hjælp været i stand til at undersøge fine enkeltheder i legemernes bygning, som vi ikke kan se med det blotte øje. Jeg nævner mikroskopet, fordi det, hvor overraskende dette end lyder, har gjort det muligt for os i dag at tælle og veje atomerne. Der er dog naturligvis ikke tale om, at vi i et mikroskop kan se de enkelte atomer og maale deres afstande. Den mindste afstand, som man paa saadan maade kan maale, er nemlig givet ved bølgelængden af det lys, som man benytter, og selv om lysbølgerne er umaadelig korte, mindre end en tusindedel af en millimeter, er de alligevel overordentlig lange sammenlignet med selve atomernes størrelse og afstande.

Imidlertid kan man i et almindeligt mikroskop iagttage smaalegemer, der, selv om de bestaar af et meget stort antal atomer, alligevel har en saa ringe vægt, at deres bevægelser, naar de er opslemmet i vand, vil paavirkes af stødene fra vandmolekylerne, der stadig rammer dem fra alle sider. Selv om smaalegemernes vægtfylde er større end vandets, vil de derfor ikke lægge sig paa bunden af vandbeholderen, men vil svømme omkring paa den mest uregelmæssige maade, dog saaledes at der til enhver tid vil være flere af dem nærmere bunden end i større højde over denne. Forholdene svarer her nøje til de usynlige bevægelser af molekylerne i luften, som jo paa grund af tyngdekraften ogsaa bliver tyndere og tyndere, jo mere vi fjerner os fra jorden. Ved at sammenligne den maade, hvorpaa tætheden aftager opefter i de to tilfælde, er det faktisk muligt at bestemme forholdet imellem vægten af et luftmolekyle og vægten af smaalegemer, der kan vejes direkte, og saa snart vi ved, hvor meget et enkelt luftmolekyle vejer, kan vi med det samme beregne, hvor mange molekyler der findes i en bestemt mængde luft.

En helt anden sindrig og i princippet simpel maade at tælle atomerne paa har man fundet ved at undersøge størrelsen af de elektriske ladninger, som smaalegemer kan optage. Saaledes kan man holde en ganske lille elektriseret oliedraabe svævende i luften ved at udsætte den for en elektrisk kraft, der trækker den opefter med akkurat samme styrke som den, hvormed tyngdekraften trækker den nedad. Naar man kender draabens vægt, faar man derfor et maal for dens elektriske ladning, og denne viser sig nu altid at være lig med et helt antal af en bestemt mindste enhed. En saadan begrænset delelighed af elektriciteten selv var man allerede meget tidligere kommet paa spor efter gennem opdagelsen af de simple love, der gælder for de mængder af grundstoffer, der udskilles ved elektrolyse af kemiske forbindelser. For samme elektricitetsmængde udskilles nemlig mange grundstoffer i mængder, der svarer til deres atomers vægte, hvis forhold man kender fra de kemiske forbindelsers sammensætning, og dette kan jo kun tydes saaledes, at den elektriske ladning, som ethvert atom i elektrolysen fører med sig, er den samme for alle saadanne stoffer. Da vi nu f. eks. efter oliedraabemetoden nøje kender enhedsladningens absolute værdi, kan vi nemt beregne antallet af atomer, der udskilles ved elektrolyse for en given elektricitetsmængde, og vejningen af den udskilte mængde giver os derfor umiddelbart vægten af det enkelte atom.

Disse store fremskridt fandt sted i det første aarti af dette aarhundrede, men allerede forinden var den udvikling begyndt, som skulde give os indgaaende oplysninger om den maade, hvorpaa legemerne er bygget op af atomer, og endda give os kendskab til atomernes indre bygning. Det første skridt var her opdagelsen af røntgenstraalerne, der har den ejendommelige gennemtrængningsevne, som bl. a. har givet lægevidenskaben nye hjælpemidler, idet det er muligt udefra at undersøge menneskers knogler og indre organer. Røntgenstraalerne har en lignende natur som lyset, men en mange gange kortere bølgelængde, og ved at undersøge straalernes tilbagekastning fra de lag, som dannes af atomerne i en krystal, har det været muligt at bestemme afstanden mellem atomerne indbyrdes og i enkeltheder at danne sig et billede af den maade, hvorpaa atomerne er ordnet i de forskellige krystaller. Ja, det har endda været muligt ved røntgenstraalernes hjælp at vinde oplysninger om, hvordan de enkelte atomer er ordnet inden for molekylerne af kemiske forbindelser.

Ved en række andre fysiske opdagelser fik man imidlertid endnu kraftigere hjælpemidler i hænde, der skulle give os kendskab til selve bestanddelene af atomerne. Et skel blev her sat, da man i elektronen lærte en fælles byggesten af alle stoffers atomer at kende. Til denne opdagelse blev man ført ved studiet af de smukke fænomener, der fremkommer ved elektriske udladninger i beholdere med stærkt fortyndet luft. I saadanne beholdere frembringes de saakaldte katodestraaler, og der, hvor disse træffer beholderens vægge, er det, at røntgenstraalerne opstaar. Da snævre bundter af katodestraaler, der kan iagttages ved deres lysende spor i luften, afbøjes i et elektrisk eller i et magnetisk felt, maatte man slutte, at disse straaler bestod af elektrisk ladede partikler, der bevæger sig med stor hastighed.

Til megen overraskelse viste det sig, at forholdet mellem katodestraalepartiklernes ladning og masse var langt større end det forhold, man finder for atomer, der udskilles ved elektrolyse, og da ladningen jo altid optræder i de samme enheder, maatte man derfor slutte, at partiklernes masse var langt mindre end atomernes masser, ja, over tusind gange mindre end massen af det letteste atom, hydrogenatomet. Det er netop katodestraalepartiklerne, som vi nu kalder for elektroner, og da disse kan frembringes ved udladninger i mange forskellige luftarter, forstod man, at elektronerne maatte findes i alle grundstoffers atomer. Længe var man imidlertid i uvidenhed om, hvad atomet ellers indeholdt, og da man var klar over, at der foruden de negativt ladede elektroner ogsaa maatte findes positive elektriske ladninger inden for atomet, maatte man først og fremmest spørge, hvordan disse ladninger var fordelt.

Vejen til at faa svar paa saadanne spørgsmaal aabnedes af den store opdagelse, at visse grundstoffer uden ydre paavirkning udsender gennemtrængende straaler. I kender jo alle den æventyrlige historie om, hvordan Madame Curie fra uranerts udvandt radium, der selv i smaa mængder er en saa kraftig straalekilde. Straalingen fra radium og de andre radioaktive stoffer har en sammensat karakter, og vi plejer at skelne imellem alfa-, beta- og gammastraaler. Medens gamma-straalerne viste sig at have lignende natur som røntgenstraalerne og betastraalerne at være elektroner med store hastigheder, fandt man, at alfa-straalerne er langt tungere partikler, der hver medfører to positive enhedsladninger og har en masse omtrentlig fire gange saa stor som hydrogenatomet. Paa grund af denne forholdsvis store masse har alfa-partiklerne saa stor bevægelsesenergi, at hver enkelt af dem, naar den standses, kan frembringe synlige virkninger. Mange af Jer har maaske selv haft den ejendommelige oplevelse at kigge i et saakaldt scintillaskop, hvor man ser et helt fyrværkeri af lysglimt, som frembringes paa en zinksulfidskærm, hvor den træffes af alfa-straalerne fra et radiumpræparat.

Disse alfa-straaler kan trænge igennem tynde lag af stoffer, men til trods for, at de paa deres vej passerer gennem millioner af atomer, er deres baner i almindelighed meget nær retliniede. Sommetider ser man imidlertid, at banen fremviser et skarpt knæk, som fortæller os, at partiklen er kommet ud for et stærkt sammenstød med noget, der er langt tungere end elektroner, der paa grund af deres lethed bare vil skubbes af vejen. Det var netop saadanne iagttagelser og betragtninger, der førte Rutherford, som har været foregangsmand paa hele dette omraade, til opdagelsen af atomkernen, hvori næsten hele atomets masse er samlet, og som samtidig er sæde for den positive elektriske ladning i atomet.

Opdagelsen af atomkernen gav os et meget simpelt billede af atomet, der paa en maade kan sammenlignes med et lille solsystem, hvori et antal lette elektroner er bundet til den tunge kerne i afstande, der langt overgaar kernens eller elektronens egen størrelse. Antallet af positive enhedsladninger paa kernen og dermed det lige saa store antal elektroner omkring denne har vist sig at være netop det samme som det tal, det saakaldte atomnummer, som angiver det paagældende grundstofs plads i det kendte system, hvori stofferne ordnes efter den ejendommelige periodiske optræden af ligheder i deres kemiske egenskaber. Den omstændighed, at disse egenskaber bestemmes af atomnummeret, fører ogsaa til en simpel forstaaelse af, at et grundstof kan optræde i forskellige afarter, som man kalder isotoper, og hvis atomkerner har samme elektriske ladning, men forskellig masse. En af de mest kendte isotoper er det tunge hydrogen, som indgaar som bestanddel i det tunge vand.

Aarsagen til grundstoffernes bestandighed ligger i, at atomkernen ikke ændres under sædvanlige kemiske omdannelser, hvor det blot er elektronerne, der omplaceres eller endda under omstændigheder helt fjernes fra kernen. Som Rutherford viste, kan dog ogsaa selve atomkernen ændres ved særlig stærk paavirkning, f. eks. ved bombardement med hurtige alfa-partikler. Derved blev for første gang muligheden af at omdanne eet grundstof til et andet paavist, og en udvikling begyndte, der skulle faa de største følger i mange retninger. Ikke alene har man gennem studiet af atomkerneomdannelser lært kernernes byggestene at kende, men gennem en rent fantastisk udvikling af tekniske hjælpemidler har man efterhaanden opnaaet et saadant kendskab til mange forskellige kernesønderdelinger, at det til slut, som alle ved, er blevet muligt i stor stil at frigøre den energi, der er bundet i atomkernerne, og som er millioner gange større end den, der kan vindes ved selv de voldsomste kemiske reaktioner.

I denne korte fortælling om lidt af det, vi har lært ved udforskning af atomernes verden, har vi hidtil kun talt om ting, der kan forstaas i simple billeder som dem, man benytter ved beskrivelser af, hvad man har set paa en rejse, men ligesom vi paa rejser i fremmede lande lærer andre skikke at kende og derved bliver klar over mange fordomme, hvormed vi drog ud, er man i atomfysikken blevet belært om en utilstrækkelighed af de principper, som fysikerne hidtil saa fast havde stolet paa.

Som jeg allerede har nævnt, minder et atom med sin atomkerne og elektroner i mange henseender om et solsystem, men en nærmere betragtning viser, at vi langtfra kan forstaa atomets egenskaber og navnlig dets særlige bestandighed ud fra de principper, hvormed Newton forklarede planeternes bevægelse omkring solen. Vel maa vi antage, at planeterne i lange tider har bevæget sig meget nær i samme baner, men dette skyldes kun at vort solsystem i saa ringe grad er blevet forstyrret af paavirkninger udefra. Skulle imidlertid engang en klode fra verdensrummet komme i nærheden af os, maa vi være forberedt paa, at baade jordens bane og aarets længde fra da af ville blive ændret. Helt anderledes forholder det sig imidlertid med atomets elektronsystem, der, saa længe kernen ikke splittes, efter enhver forstyrrelse vil reorganiseres saaledes, at den oprindelige tilstand igen vender tilbage.

En nøgle til forstaaelsen af de for atomerne gældende ejendommelige lovmæssigheder har vi faaet igennem Plancks opdagelse i det første aar af vort aarhundrede af det saakaldte virkningskvantum, der er udtryk for en begrænsning af selve de mekaniske processers delelighed. Det er dog slet ikke muligt at forklare en saadan begrænsning ved hjælp af vore sædvanlige fysiske begreber, der, hvor udstrakt deres anvendelsesomraade end er, har vist sig at være idealisationer, der kun gælder, naar de virkninger, det drejer sig om, er store i forhold til det enkelte kvantum, og som svigter, naar vi søger at anvende dem til at gøre rede for de nye erfaringer fra atomernes verden.

Nogle af Jer har maaske allerede hørt om de overraskende vanskeligheder, som spørgsmaalet om elektronernes natur har stillet os overfor. Paa den ene side maa vi jo betragte elektronen som en partikel, idet maalinger af en elektrons masse og elektriske ladning altid giver samme resultater. Paa den anden side er man ved beskrivelsen af andre af elektronernes egenskaber henvist til at bruge bølgebilleder i lighed med dem, der har vist sig uundværlige for beskrivelse af lysets forplantning. Overraskelser af samme art har vi ogsaa mødt ved spørgsmaalet om lysets natur, idet bølgebillederne er ganske utilstrækkelige for at forklare love, der gælder for den maade, hvorpaa atomerne optager og udsender lysenergi, og for hvis redegørelse man har maattet gribe til at beskrive lyset som sammensat af enkelte lyskvanter eller fotoner med partikelkarakter.

En saadan situation, der hidtil var helt ukendt i fysikken, maatte jo til at begynde med virke ganske forvirrende, men efterhaanden indsaa man, at naar man benyttede de modstridende billeder, var det aldrig for at beskrive eet og samme fænomen, men kun for at gøre rede for erfaringer, der var vundet under forskellige forsøgsbetingelser, der udelukkede hinanden. Saadanne erfaringer staar derfor i et forhold til hverandre, som man betegner som **komplementært** for at understrege, at de, til trods for at de ikke kan forenes i eet enkelt anskueligt billede, hver især giver Udtryk for lige vigtige sider af de samlede oplysninger, der overhovedet kan vindes.

I denne forbindelse maa vi først og fremmest gøre os klart, at vi for at beskrive en bestemt forsøgsopstilling og de derved vundne iagttagelsesresultater altid vil være henvist til at benytte begreber udtrykt i den sprogbrug, som menneskene har udviklet for at finde sig til rette ved dagliglivets begivenheder. Dette betyder jo ikke andet end, at det ved ethvert fysisk forsøg maa dreje sig om noget, hvorom vi kan fortælle andre, hvad vi har gjort, og hvad vi har lært. Helt anderledes stiller spørgsmaalet sig imidlertid, naar vi skal sammenfatte resultater, vundet under forskellige forsøgsbetingelser.

Kun saa længe, det drejer sig om fænomener, hvor vi kan se helt bort fra den vekselvirkning, der altid maa være til stede mellem genstanden for undersøgelsen og de apparater, der benyttes ved dens iagttagelse, som f. eks. naar vi bestemmer en planets sted i dens bane til forskellige tider, kan vi tale om genstandens selvstændige opførsel uafhængig af vor iagttagelse. Naar vi derimod har at gøre med en elektron under forhold, hvor virkningskvantet spiller en afgørende rolle, kan vi slet ikke længere skelne skarpt imellem, hvad vi kunne kalde elektronens egen opførsel og dens vekselvirkning med de maaleinstrumenter, der indgaar i den forsøgsopstilling, hvorunder fænomenet fremkommer. Under saadanne forhold er derfor forudsætningen for paa sædvanlig maade at sammenligne resultater vundet under forskellige forsøgsbetingelser ikke til stede. Det drejer sig her ingenlunde om vanskeligheder af teknisk art, men om et principielt punkt, der er betinget af selve vilkaarene for erfaringernes beskrivelse.

Den komplementære sammenfatningsmaade er at betragte som en naturlig udvidelse af den sædvanlige fysiske beskrivelse, der ikke har vist sig rummelig nok til at omfatte de ejendommelige lovmæssigheder, der gælder i atomernes verden, og som er afgørende for egenskaberne af de stoffer, hvoraf alle vore redskaber, ja ogsaa vore egne legemer bestaar. I første øjeblik lyder det maaske underligt, at man skulle være nødt til at give afkald paa de krav om anskuelighed i naturbeskrivelsen, som man hidtil antog som helt selvfølgelige, men ved nærmere betragtning finder man, at vi ogsaa paa mange andre omraader har at gøre med forhold, der viser en nøje lighed med dem, vi har mødt ved undersøgelsen af atomernes egenskaber.

Vi behøver jo blot at tænke paa den komplementære maade, hvorpaa vi bruger saadanne to ord som „tanker" og „følelser" for at beskrive de situationer, som ethvert menneske daglig befinder sig i. Disse ord henviser jo netop til sider af vore indre oplevelser, der er lige væsentlige, men som udelukker hinanden i den forstand, at selv vore varmeste følelser fuldstændig mister deres art, naar vi forsøger at udrede dem ad den kolde logiske tankes vej. Lignende forhold gør sig gældende i samlivet med vore medmennesker, hvor intet af ordene „retfærdighed" og „kærlighed" kan undværes. Enhver pige eller dreng forstaar, hvor meget „fair play" betyder i leg eller sport, og at man ingen agtelse kan have for sig selv, uden naar man stræber efter at handle retsindigt mod andre, men samtidig maa vi gøre os klart, at brugen af retfærdighedsbegrebet i dets yderste konsekvens udelukker den kærlighed, som kaldes paa over for forældre, søskende og kammerater.

Naar man først har indlevet sig deri, kan man efterspore komplementaritetsforhold paa saa at sige alle kulturlivets omraader, og den omstændighed, at vi tvinges til at erkende saadanne forhold inden for et omraade som fysikken, der paa grund af dens anvendelser indtager saa stor en plads i skoleundervisningen, turde ogsaa komme til at øve en indflydelse paa de unges opdragelse. Selv om det i et kort foredrag ikke er muligt at give en klar og indgaaende fremstilling hverken af den store udvikling inden for atomfysikken eller af den belæring om vilkaarene for vor erkendelse, den har bragt, haaber jeg, at det er lykkedes mig i det mindste at give Jer et indtryk af problemernes betydning og rækkevidde.

TRANSLATION

Niels Bohr:

Atoms and Human Knowledge

Upon request Professor Bohr has given us the manuscript for the talk he gave on the school radio yesterday for Danish and Norwegian gymnasium students

It is a great pleasure for me to have such an opportunity to come into contact with the young people in Denmark and Norway who through education in gymnasium prepare themselves for a career in the service of society. The title of my talk, "atoms and human knowledge", may sound very learned, but I shall try to tell a little about the insight into the world of atoms, gained by one of the most fantastic journeys of the human spirit into unknown reaches, where we have perpetually met great surprises and have had to overcome many difficulties. The experience we have gathered on this journey signifies a great advance in our mastery of nature, but at the same time we have gained a general human lesson which holds a message for everyone, both old and young.

The idea that all bodies we see or touch are built up of atoms, which with their tiny size evade direct observation, goes far back in time, and Greek philosophers such as Democritus already understood that in order to explain the stability of the elements it was necessary to assume a limit for the divisibility of substances. When, for example, ordinary water can be changed by vaporization into a completely different state and the steam can then in turn be condensed to water with exactly the same properties as before, this would not be possible to understand if we had to do with complete transformation of the substances, but the situation can be explained quite simply when we assume that what happens at vaporization is just that the tightly-packed small parts in the water are separated from each other so that they for a time can move more freely among one another, and that on condensation nothing more happens than that the molecules are again packed together and their mobility is limited.

As time passed, such ideas proved to be extremely fruitful and enlightening in many different fields, and especially at the beginning of the last century we gained through Dalton's work a simple explanation of the constant mass ratios in which elements combine into chemical compounds. We assume of course that in an element all atoms are identical while in a chemical compound there are small parts or molecules made up of atoms of various elements. For example, water is a combination of the elements *brint* and *ilt*, as they are called in Danish or *vandstof* and *surstof* in Norwegian, or, in terms which are used in so many other places and which therefore are also in the process of being introduced in Denmark: hydrogen and oxygen. A water molecule consists of two atoms of hydrogen and one atom of oxygen, but there is also another combination of these elements, hydrogen peroxide, whose molecules consist of two hydrogen atoms and two oxygen atoms, and on the formation of this substance the same amount of hydrogen will therefore combine with precisely double the amount of oxygen as on the formation of water.

Even though in many such ways the theory of atoms proved to be ever more indispensable and convincing, it was nevertheless the general view up to our century that it would never be possible to give a direct proof of the existence of atoms, as it was assumed that the coarseness of our senses would exclude any demonstration of the presence and the effect of single atoms. In the course of the development of physics, however, we have obtained more and more refined instruments to assist our senses, and we have for example long since been able, by means of the microscope, to study the fine details of the structure of bodies that we cannot see with the naked eye. I mention the microscope because, however surprising it may sound, it has made it possible for us today to count and weigh atoms. Still, this naturally does not mean that we in the microscope can see individual atoms and measure the distances between them. The smallest distance we can measure in this way is given by the wavelength of the light we use, and although light waves are extremely short, less than one thousandth of a millimetre, they are nevertheless exceedingly long as compared to the size and distances of the atoms.

However, in an ordinary microscope we can observe small bodies which, although they consist of a very large number of atoms, nevertheless have so little weight that their movements, when they are suspended in water, will be affected by the impact from the water molecules which continually hit them from all sides. Although the density of the small bodies is higher than that of water, they will not therefore fall to the bottom of the water vessel but swim about in the most irregular way, though in a manner such that at any time more of them will be nearer the bottom than at some height above it. This situation corresponds closely to the invisible movements of the molecules in the air which due to gravity also becomes thinner and thinner the further away we are from Earth. By comparing the way in which the density decreases in upward direction in these two cases, it is in fact possible to determine the ratio between the weight of an air molecule and the weight of small bodies that can be weighed directly, and as soon as we know how much a single air molecule weighs, we can immediately calculate how many molecules are present in a given amount of air.

A completely different ingenious, and in principle simple, way of counting atoms has been found by examining the magnitude of the electric charges that small bodies can assume. Thus it is possible to keep a quite small electrified drop of oil floating in the air by subjecting it to an electric charge, which pulls it upwards with exactly the same force as gravity pulls it downwards. When the weight of the drop is known, a measure of its electric charge is therefore obtained, and this now always proves to be equal to a whole number of a certain smallest unit. Such a limited divisibility of electricity itself was recognized already much earlier through the discovery of the simple laws that hold for the amounts of elements that are separated out by electrolysis of chemical compounds. Thus, for the same quantity of electricity many elements are separated out in amounts corresponding to the weight of their atoms, whose ratios are known from the composition of the chemical compounds, and this can of course only be interpreted to mean that the electrical charge that each atom carries with it in the electrolysis is the same for all such substances. As we now know exactly, for example from the oil drop method, the absolute value of the elementary charge, we can easily calculate the number of atoms separated out by electrolysis for a given electric charge, and the weighing of the separated amount thus provides us directly with the weight of the individual atom.

These great advances took place in the first decade of this century, but already before this the development was begun which was to provide us with detailed information about the manner in which the bodies are built up of atoms, and even give us knowledge about the internal constitution of atoms. Here, the first step was the discovery of X-rays, which have the peculiar ability of penetration that among other things has provided medical science with new tools, in that it is possible to study the bones and inner organs of the human body from the outside. X-rays are of a nature similar to that of light but have a wavelength which is many times shorter, and by studying the reflection of the rays from the layers formed by the atoms in a crystal, it has been possible to determine the distance between the atoms and to form a detailed picture of the way in which the atoms are arranged in various crystals. Indeed, it has even been possible by means of X-rays to gain information about how the individual atoms are placed inside the molecules of chemical compounds.

Through a series of other discoveries in physics, however, we got still more powerful tools in our hands which would give us knowledge about the very components of the atoms. This was a milestone, since with the electron we became acquainted with a constituent common to the atoms of all substances. We were guided to this discovery through the study of the beautiful phenomena which appear in connection with electric discharges in containers with extremely rarefied air. In such containers we can produce the so-called cathode rays, and where these hit the walls of the container, X-rays are produced. Since narrow bundles of cathode rays, which can be observed due to their luminous tracks in air, are deflected in an electric or in a magnetic field, it could be concluded that these rays consisted of electrically charged particles moving at high speed.

Very surprisingly, it turned out that the ratio between the charge and the mass of the cathode-ray particles was much higher than the ratio found in atoms separated out in electrolysis, and since charge always appears in the same units, it was thus necessary to conclude that the mass of the particles was far smaller than the mass of the atoms, indeed more than a thousand times smaller than the mass of the lightest atom, the hydrogen atom. It is precisely the cathode ray particles that we now call electrons, and since they can be produced in discharges in many different gases, it was understood that electrons must be present in the atoms of all elements. For a long time, however, it was not known what else was inside the atom and, as it was clear that in addition to the negatively charged electrons there must also be positive electric charges inside the atom, the question to be asked first of all was how these charges were distributed.

The path to an answer to such questions was opened by the great discovery that certain elements emit penetrating radiation even without external influence. You all know the fantastic story of how Madame Curie extracted radium, which even in small quantities is such a strong source of radiation, from uranium ore. The radiation from radium and other radioactive substances is of a complex nature, and we usually distinguish between alpha, beta and gamma rays. Whereas the gamma rays proved to have a nature similar to X-rays, and beta rays to be electrons at high velocities, it was found that alpha rays are much heavier particles, each of which carrying two positive unit charges and having a mass about four times that of the hydrogen atom. Because of this relatively large mass, alpha particles have so great a kinetic energy that every single one of them, as it is stopped, can produce visible effects. Many of you have perhaps had the strange experience of looking into a so-called scintilloscope where one can see a whole fireworks of light flashes produced on a zink sulphide screen where it is hit by the alpha rays from a radium preparation. These alpha rays can penetrate thin layers of material, but although they pass through millions of atoms on their way, their paths are usually almost straight lines. Sometimes, however, one sees that the path shows a sharp break, which tells us that the particle has had a violent collision with something much heavier than electrons which, due to their lightness, will just be pushed aside. Just such observations and considerations led Rutherford, who has been a pioneer in this entire field, to the discovery of the atomic nucleus where almost the entire mass of the atom is concentrated and which at the same time is the seat of the positive electric charge in the atom.

The discovery of the atomic nucleus gave us a very simple picture of the atom, which in a way can be compared to a small solar system where a number of light electrons are bound to the heavy nucleus at distances far surpassing the size of the nucleus or the electrons themselves. The number of positive unit charges on the nucleus, and hence the equal number of electrons around it, has turned out to be just the same as the number, the so-called atomic number, denoting this element's place in the familiar system where the elements are ordered according to the peculiar periodic appearance of similarities in their chemical properties. The circumstance that these properties are determined by the atomic number also leads to a simple understanding of the fact that an element can appear in different variants, called isotopes, whose atomic nuclei have the same electric charge but different mass. One of the best known isotopes is heavy hydrogen which is a constituent of heavy water.

The reason for the stability of the elements is that the atomic nucleus is not changed during ordinary chemical transformations in which it is only the electrons that are rearranged, or, under certain circumstances, are even entirely removed from the nucleus. As Rutherford showed, also the atomic nucleus itself, however, can be changed by especially strong influence, for example during bombardment with fast alpha particles. In this way the possibility of transforming one element into another was demonstrated for the first time, and a development began which would have the greatest consequences in many domains. Not only have we learned about the constituents of the nuclei through the study of the transformations of atomic nuclei, but through a quite fantastic development of technical tools we have gradually gained so much information about many different nuclear disintegrations that it has finally, as everybody knows, become possible to release on a large scale the energy bound in atomic nuclei, which is millions of times greater than that which can be gained in even the most violent chemical reactions.

In this brief description about a little of what we have learnt from investigating the world of atoms, we have hitherto only spoken of matters that can be understood in terms of simple pictures such as those we use when describing what we have seen on a journey, but just as we on journeys in foreign countries learn about other customs and thus become aware of many prejudices that we took with us, we have been taught in atomic physics about the insufficiency of the principles on which the physicists had hitherto relied so completely.

As I have mentioned already, an atom with its nucleus and electrons resembles in many respects a solar system, but a closer look shows that we are far from understanding the properties of the atom, and its special stability in particular, on the basis of the principles by which Newton explained the movements of the planets around the sun. We surely may assume that the planets for a long time have moved very nearly in the same orbits, but this is only because our solar system to such a small extent has been disturbed by outside influences. However, should at some time a body from outer space come close to us, we must be prepared that both the orbit of the earth and the length of the year will be changed from then on. The situation is quite different, however, with regard to the electron system of the atom, which as long as the nucleus is not split will be reorganized after every disturbance, so that the original state returns once again.

We have obtained a key to the understanding of the strange regularities valid for atoms through Planck's discovery in the first year of our century of the so-called quantum of action, which is an expression of a limitation of the divisibility of the mechanical processes themselves. It is, though, not at all possible to explain such a limitation by means of our usual physical concepts which, however broad their area of application might be, have proved to be idealizations which are only valid when the effects in question are large in relation to the individual quantum and which fail when we try to use them in order to explain the new experiences from the world of atoms.

Some of you have perhaps already heard about the surprising difficulties with which the question of the nature of electrons has confronted us. On the one hand we may of course regard the electron as a particle, since measurements of the mass and the electric charge of an electron always give the same results. On the other hand, when describing other properties of the electrons we have to use wave pictures similar to those that have proved indispensable for the description of the propagation of light. We have also met surprises of the same kind concerning the question of the nature of light, since the wave pictures are quite insufficient to explain laws valid for the way in which atoms absorb and emit light energy, and to explain this we have had to resort to a description of light as composed of individual light quanta or photons behaving as particles.

Such a situation, which was hitherto quite unknown in physics, must to begin with seem quite confusing, but gradually it was realized that when the contradictory pictures were used, it was never to describe one and the same phenomenon, but only to explain experiences gained under different experimental conditions excluding one another. Such experiences are therefore in a relation to each other that we call *complementary* in order to emphasize that although they cannot be united in one single visualizable picture, each of them is an expression of equally important aspects of the total information that can be gained at all.

In this connection we must first of all be aware that in order to describe a given experimental arrangement and the observation results thereby obtained, we will always have to employ concepts expressed in the language usage that people have developed in order to cope with everyday events. This means nothing more than that every physical experiment must concern something about which we can tell others what we have done and what we have learnt. The situation is quite different, however, when our purpose is to comprehend results obtained under different experimental conditions.

Only as long as we have to do with phenomena where we can entirely ignore the interaction which must always exist between the object of investigation and the apparatus used in its observation, as for example when we determine a planet's position in its orbit at different times, can we speak of the object's independent behaviour regardless of our observation. When, on the contrary, we have to do with an electron under conditions where the quantum of action plays a decisive role, we cannot any longer distinguish sharply between what we might call the electron's own behaviour and its interaction with the measuring instruments which are part of the experimental arrangement in which the phenomenon occurs. Therefore, under such circumstances the prerequisite for comparing in the usual manner the results obtained under different experimental conditions is no longer present. Here we do not at all have to do with difficulties of a technical kind but with a fundamental point which depends on the very conditions for the description of the experiences.

The complementary approach to comprehension is to be regarded as a natural extension of the usual physical description, which has not proved sufficient to accommodate the strange regularities valid in the world of atoms and decisive for the properties of the substances of which all our tools, indeed also our own bodies, consist. At first it may sound surprising that we should be forced to renounce the requirements for visualizability in the description of nature that was hitherto assumed to be quite self-evident, but by closer consideration we find that also in many other domains we have to do with circumstances that show a close similarity with those we have met in the study of the properties of atoms.

We need only to think of the complementary way in which we use two such words as "thoughts" and "feelings" in order to describe the situations in which every human being finds itself every day. These words refer precisely to aspects of our inner experiences, each of which is equally important but excludes the other in the sense that even our warmest feelings completely lose their character when we try to clarify them along the path of cold logical thought. Similar conditions hold in our lives together with our fellow beings in which neither of the words "justice" and "love" can be dispensed with. Every girl or boy understands how much "fair play" means in games or sport and that one cannot have respect for oneself unless one tries to act honourably towards others, but at the same time we must be aware that the use of the concept of justice in its utmost consequence excludes the love which is called upon in relation to parents, sisters, brothers and comrades.

When first one has become familiar with this, it is possible to trace complementarity in so to speak all spheres of cultural life, and the circumstance that we are forced to recognize such situations within a field such as physics, which because of its applications occupies such a large place in the school curriculum, may be assumed to come to have an influence on the upbringing of young people. Although in a short talk it is not possible to give a clear and detailed exposition of either the great development within atomic physics or the lesson regarding the conditions for human knowledge this has brought, I hope that I have at least succeeded in giving you an impression of the importance and extent of the problems.

VOLUME 12, PART II

PEOPLE

Niels Bohr and Ernest Rutherford, Cambridge, 1923.

V. PROFESSOR SIR ERNEST RUTHERFORD AND HIS SIGNIFICANCE FOR THE RECENT DEVELOPMENT OF PHYSICS

PROFESSOR SIR ERNEST RUTHERFORD OG HANS BETYDNING FOR FYSIKENS NYERE UDVIKLING

Politiken, 18 September 1920, pp. 7–8

TEXT AND TRANSLATION

«Politiken»s Kronik.

Professor Sir Ernest Rutherford og hans Betydning for Fysikens nyere Udvikling.

Den verdensberømte engelske Fysiker, Professor Sir Ernest Rutherford begynder i Dag paa vort Universitet en kort Række Forelæsninger efter Indbydelse af Danmarks naturvidenskabelige Samfund. Nedenfor giver hans danske Kollega Professor Niels Bohr en Skildring af Rutherford og hans Betydning for Fysikens nyere Udvikling.

Den Mand, hvis Navn staar over disse Linjer, og som i disse Dage gæster Kjøbenhavn for efter Indbydelse af Danmarks naturvidenskabelige Samfund at holde en Række Forelæsninger paa Universitetet, har bag sig et videnskabeligt Livsværk, som selv paa Baggrund af den vidunderlige Udvikling, som Naturvidenskaben har undergaaet i de sidste Decennier, tegner sig paa den mest straalende Maade.

For blot 20—30 Aar tilbage var det en almindelig udbredt Anskuelse, at den fysiske Videnskab maatte søge sine Fremtidsmuligheder igennem Forfinelse af Maalingerne af de kendte fysiske Fænomener og af den matematiske Behandling af Maalingsresultaterne, idet man mente sig berettiget til den Antagelse, at man kunde overse Omraadet af iagttagelige fysiske Fænomener og de for dem gældende Naturlove. Vel var man klar over, at de iagttagne Fænomener i og for sig ikke var af primær Natur, og at de fremkom som Sammenspil af Virkninger af det uhyre Antal Smaadele, hvoraf de iagttagelige Legemer maatte tænkes at bestaa; men man følte sig overbevist om, at det paa Grund af disse Smaadeles Lidenhed stadig vilde ligge uden for den af Menneskene opnaaelige Grænse at iagttage Virkningerne af de enkelte Smaadele hver for sig. Vidnesbyrd om dette Synspunkt finder man i Lærebøger i Fysik og Kemi fra den omhandlede Tid, hvor det stadig fremhæves, at man til Trods for Atomhypotesens tilsyneladende Nødvendighed for Beskrivelsen af Naturfænomener, dog altid maatte holde for Øje, at man havde med en saakaldt Arbejdshypotese at gøre, selv om det af alle maatte indrømmes,at det her drejede sig om en saadan af enestaaendeFrugtbarhed. Dette Forhold har imidlertid fuldstændig forandret sig i den forløbne Tid, idet vi nu om Stunder ikke alene har uomtvistelige Beviser paa de enkelteAtomers Eksistens, ja behersker Metoder til med stor Nøjagtighed at bestemme Antallet af Atomer i de iagttagelige Legemer, men endda besidder et indgaaende Kendskab til de Byggestene, hvoraf de forskellige Grundstoffers Atomer er sammensat.

Denne Udvikling er fremkommet ved Sammenvirkning af mange forskellige Omstændigheder. Dels er man igennem teoretiske Studier ført til nye Synspunkter, der har aabenbaret en uanet Sammenhæng mellem tilsyneladende vidt forskellige Grupper af Fænomener, der, idet de maa tilskrives Virkninger af de samme Smaadele, sammenholdt aabner Mulighed for at drage bestemte Slutninger om de enkelte Smaadeles Virkninger. Som et Resultat, naaet paa denne Maade, kan nævnes, at det har været muligt ved Sammenligning mellem Resultater, hentet fra den mekaniske Varmeteori og den elektromagnetiske Varmestraalingsteori med stor Nøjagtighed at bestemme Antallet af Atomer i et Gram af et givet Grundstof; for et Gram Brint er dette Tal cirka seks Hundred Tusind Milliard Milliarder. Desuden har man lært et Antal nye Fænomener at kende, der, om end betinget af Tilstedeværelsen af de sædvanlige, kendte Stoffer, i Modsætning til de tidligere kendte Fænomener giver os Vidnesbyrd af mere direkte Karakter om de enkelte Atomers Virkninger. Dette gælder navnlig de mange smukke Fænomener, som man har opdaget ved det nærmere Studium af elektriske Udladninger gennem Luftarter. Jeg skal blot nævne, at Sir J. J. Thomson, hvis banebrydende Arbejder vi først og fremmest skylder vort Kendskab til og Forstaaelse af disse Fænomener, paa Basis af et Studium af disse ikke alene har kunnet uddanne en forholdsvis direkte Metode til Bestemmelse af Atomernes Antal, der har givet Resultater, der stemmer nøje overens med dem, der udledes ad den nys omtalte, mere indirekte teoretiske Vej, men ogsaa har kunnet fastslaa Tilstedeværelsen i alle Grundstoffers Atomer af smaa, ensartede Partikler, de saakaldte *Elektroner*, der besidder en negativ elektrisk Ladning og en Masse, som er meget lille i Forhold til hele Atomets Masse. Dette sidste Resultat har haft den største Betydning, idet man saaledes for første Gang kunde iagttage og undersøge i alt Fald visse af Atomernes Byggestene.

Disse Resultater skulde imidlertid hurtigt blive fuldstændiggjorte og overfløjne ved Opdagelsen og Undersøgelsen af et Antal Grundstoffer med helt nye Egenskaber, nemlig de saakaldte *radioaktive* Stoffer. Mange Læsere vil kunne erindre, hvor stor Interesse Opdagelsen af Radium, der navnlig skyldtes Mme. Curie's indsigtsfulde og energiske Arbejde, vakte omkring Aarhundredskiftet, og den Opsigt, de vidunderlige og tilsyneladende ubegribelige Virkninger af dette Stof fremkaldte. I Øjeblikket staar disse Virkninger dog ikke mere i et saadant mystisk Skær som den Gang, idet der gennem det enestaaende geniale Arbejde, som Sir *Ernest Rutherford* i Løbet af den siden da forløbne Tid har udført, er kastet det klarest tænkelige Lys over de fleste af de Fænomener, der knytter sig til de radioaktive Stoffer. Hans Studier af disse Fænomener har ledet til Resultater, der ikke alene ikke strider imod de Slutninger, man havde draget af tidligere Erfaringer, men netop belyser og tillader en dybere Forstaaelse af disse, samtidig med at de har aabnet helt nye og uanede Veje til at trænge ind i Spørgsmaalet om Atomernes Bygning.

Rutherfords første store Indsats paa detteOmraade bestod deri, at han erkendte, at den tilsyneladende lunefulde Optræden og Foranderlighed af de iagttagne Fænomener fuldkommen forsvandt, dersom man antog, at man i de radioaktive Stoffer havde at gøre med Grundstoffer, hvis Atomer uden ydre Indgreb undergik eksplosionsagtige Forandringer, ved hvilke elektrisk ladede Dele af Atomet, de saakaldte A(alfa)- og B(beta)-Partikler, udslyngedes, samtidig med at der af det tilbageblevne dannedes et Atom af et nyt Grundstof. Medens det nye Stof i mange Tilfælde vil være af en lige saa stabil Natur som de sædvanlige Grundstoffer, vil det ofte selv igen være radio-aktivt og dets Atomer undergaa en ny eksplosionsagtig Forvandling o. s. v. For denne dristige Hypotese, der med et Slag bragte Orden og Sammenhæng i det tilsyneladende Virvar, lykkedes det Rutherford og hans Medarbejdere at faa afgørende Bevis ved at isolere et stort Antal saadanne Forvandlingsprodukter, hvis kemiske Undersøgelse viste, at de besad Egenskaber, der ganske svarede til de kendte Grundstoffer, fra hvilke de kun skilte sig ved, at deres Atomer havde en begrænset „Levetid", der før eller senere afbrydes af en Eksplosion.

I de følgende Aar undersøgte Rutherford igennem en Række storslaaet anlagte og glimrende udførte Forsøg de Fænomener, som ledsager Atomernes Eksplosion. Som et særlig smukt Resultat af disse Undersøgelser kan nævnes, at det viste sig, at de ved Eksplosionen udsendte Partikler besad saa stor Bevægelsesenergi, at Virkningen af en enkelt udslynget Partikel kunde iagttages. Herved kom man i Besiddelse af en direkte Metode til at tælle Atomerne, hvis Resultater viste sig at stemme fuldkommen overens med dem, der var fundet ved Hjælp af de før omtalte Metoder. Det Hovedresultat, Rutherford opnaaede, var dog de Slutninger, han formaaede at drage om Atomernes Byggestene. Idet de enkelte Partiklers Baner kunde følges paa deres Gang gennem Luftarter og andre Stoffer, viste det sig nemlig, at man kunde faa Oplysning om Bygningen af de Atomer, de mødte. Det fandtes saaledes, at et Atom i Almindelighed var meget let gennemtrængelig for Partiklerne, idet disse oftest uden mærkelig Retningsforandring kunde gennemtrænge mangeTusinde Atomer, Men en enkelt Gang fandtes Partiklen ved Sammenstødet med et Atom at lide storAfbøjning, hvilket, som Rutherford viste, maatte tydes derved, at Atomet foruden de negativ ladede Elektroner besad en Kerne, som var ladet med positiv Elektricitet, og som indeholdt næsten hele Atomets Masse. Da Kernen er uhyre lille i Forhold til hele Atomet, er dette bygget paa ganske lignende Maade som et Solsystem; Atomet bestaar af et Antal lette Elektroner, der bevæger sig omkring den tunge Atomkerne, som Planeterne omkring Solen. Ved denne Rutherford's Opdagelse af *Atomkernen*, der blev bestyrket paa mangfoldige Maader, maa man antage at have lært samtlige Atomets Byggestene at kende, og man besidder derfor nu et Grundlag for at trænge ind i Spørgsmaalet om Forklaringen af Stoffernes Egenskaber ved Hjælp af nøjere udarbejdede Forestillinger om Atombygningen.

Medens det ikke vil være muligt her at komme nærmere ind paa den teoretiske Behandling af det sidste Spørgsmaal og de forjættende Udsigter, der aabner sig for at skabe en fælles Basis for Forklaringen af saa vel Kemiens som Fysikens Fænomener, skal jeg endnu blot omtale det nye vigtige Resultat, som Rutherford har naaet ved Undersøgelser i den allersidste Tid. Ved overordentlig fint udførte og vanskelige Eksperimenter har han kunnet iagttage, hvorledes ved et Sammenstød mellem et Atom og en Alfa-Partikel, hvor Partiklen rammer lige ind i Atomkernen — hvad naturligvis sker uhyre sjældent — denne sidste under Omstændigheder kan sønderdeles i Brudstykker, der hver for sig vil danne Kerner for Atomer af andre Grundstoffer. Idet man gennem dette Resultat har fundet Midler til vilkaarlig at fremkalde en Proces, der ganske svarer til den, der spontant foregaar i den ved de radio-aktive Stoffers Forvandlinger iagttagne Eksplosioner af Atomkernen, kan Naturvidenskaben for saa vidt siges at være indtraadt i en ny Epoke, som Bestandigheden af Grundstoffernes Atomer overfor ydre Angreb hidtil har været anset for den faste Grund for fysiske og kemiske Betragtninger. Disse Linjers Hovedformaal har dog netop været at forsøge at give Læseren Indtryk af, hvor naturligt og direkte Rutherfords nye Resultat slutter sig til hans tidligere ovenfor nævnte Resultater, hvorigennem de Grundforestillinger, hvormed de kemiske og fysiske Videnskaber arbejder, er udviklet til en vidunderlig og uanet Klarhed.

Det ovenstaaende er naturligvis kun en kort Skitse af ganske faa Hovedresultater af Sir Ernest Rutherfords Arbejder, der ved Omfang og Idérigdom er uden Sidestykke i Nutidens eksperimentalfysiske Forskning. Naar det har været muligt for en enkelt Mand at gøre en saa enestaaende Indsats i Videnskaben, skyldes det naturligvis først og fremmest hans overlegne videnskabelige Geni, men samtidig tillige en Række almenmenneskelige Egenskaber, der har sat ham i Stand til at samle et stort Antal Elever omkring sig og vejlede og begejstre dem. Det ligger uden for min Evne blot at forsøge at beskrive disse Egenskaber nærmere, men dette er jo ogsaa saa meget mindre nødvendigt, som den danske videnskabeligt interesserede Offentlighed i disse Dage vil faa Lejlighed til selv at lære hans karakterfulde Personlighed at kende. Som en af de mange, der har haft den Lykke at arbejde i hans Laboratorium under hans Ledelse, vil jeg blot gerne give Udtryk for den Taknemmelighed og Beundring, hvormed Sir Ernest Rutherford omfattes af alle hans Elever.

N. Bohr.

TRANSLATION

Professor Sir Ernest Rutherford and his Significance for the Recent Development of Physics.

Professor Sir Ernest Rutherford, the world-famous English physicist, starts today at our University a short series of lectures at the invitation of the Danish Science Society. Below, his Danish colleague Professor Niels Bohr gives a description of Rutherford and his significance for the recent development of physics.

The man whose name stands above these lines and who is visiting Copenhagen at the present time on the invitation of the Danish Science Society in order to give a series of lectures at the University, has behind him a life in science which, even on the background of the wonderful development that science has gone through in the last decades, appears in the most brilliant fashion.

Only 20–30 years ago it was generally assumed that the future of physical science was to be sought through the refinement of the measurements of the known physical phenomena and of the mathematical treatment of the measurement results, in that one felt entitled to assume that one could embrace the area of observable physical phenomena and the laws of nature appertaining to them. It was clear, of course, that the observed phenomena on the whole were not of a primary nature, and that they arose as a combination of effects of the enormous number of small parts that observable bodies must be thought to consist of; but one felt convinced that because of the minuteness of these small parts it would remain beyond the limits of humans to observe the effects of the individual small parts one by one. Evidence for this viewpoint is to be found in physics and chemistry textbooks from the time in question, where it is constantly emphasized that despite the seeming necessity of the atomic hypothesis for the description of natural phenomena, one should nevertheless always keep in mind that one had to do with a so-called working hypothesis, even though this must be admitted by all to be one of singular fruitfulness. However, this state of affairs has changed completely in the meantime, as we nowadays not only have irrefutable evidence for the existence of single atoms – indeed master methods for determining the number of atoms in observed bodies with great accuracy – but even have detailed knowledge about the building blocks which make up the atoms of the various elements.

This development has come about through the interplay between many different circumstances. Partly, one is led through theoretical studies to new viewpoints that have revealed an unsuspected relationship between seemingly completely different groups of phenomena which, as they must be ascribed to effects of the same small parts, taken together open the possibility of drawing definite conclusions about the effects of the individual small parts. As a result reached in this way can be mentioned that it has been possible by comparison of results taken from the mechanical theory of heat and the electromagnetic theory of heat radiation to determine with great accuracy the number of atoms in one gramme of a known element; for one gramme of hydrogen this number is about six hundred thousand milliard milliards. In addition, a number of new phenomena have become known, which, although contingent on the presence of the usual known substances, contrary to previously known phenomena, provide evidence of a more direct character about the effects of single atoms. This holds in particular for the many beautiful phenomena which have been discovered by the closer study of electric discharges in gases. I will only mention that Sir J.J. Thomson, to whose pioneering publications we first and foremost owe our knowledge and understanding of these phenomena, has on the basis of a study of these not only been able to develop a relatively direct method for determining of the number of atoms, which has produced results agreeing exactly with those derived in the more indirect way just discussed, but has also been able to determine, in the atoms of all elements, the presence of small identical particles, the so-called *electrons*, which have a negative electrical charge and a mass which is very small in relation to the mass of the whole atom. This last result has had the greatest significance in that it has been possible for the first time to observe and investigate at least some of the building blocks of the atoms.

These results should, however, quickly be rounded off and surpassed by the discovery and investigation of a number of elements with entirely new properties, namely the so-called *radioactive* elements. Many readers will remember how great an interest the discovery of radium, which was due in particular to Mme Curie's perspicacious and energetic work, aroused at the turn of the century, and the sensation created by the wonderful and seemingly incomprehensible effects of this substance. At present, though, these effects are no longer surrounded by such a mysterious aura as at that time, since through the unique brilliant work performed by Sir *Ernest Rutherford* during the time elapsed since then, the clearest imaginable light has been thrown on most of the phenomena which are linked to radioactive substances. His studies of these phenomena have led to results which not only do not conflict with the conclusions drawn from earlier experience, but precisely illuminate and allow a deeper understanding of them, while they have opened quite new and unimagined ways to penetrate more deeply into the question of the constitution of atoms.

Rutherford's first great success in this field consisted of his recognition that the seemingly capricious behaviour and variability of the observed phenomena disappeared completely if it was assumed that in dealing with radioactive substances one had to do with elements whose atoms, without external influence, underwent explosion-like changes, by which electrically charged parts of the atom, the so-called A(alpha) and B(beta) particles, were ejected, while an atom of a new element was formed from what was left behind. Whereas in many cases the new element will be of just as stable a nature as the ordinary elements, it will often itself also be radioactive and its atoms will suffer a new explosion-like transmutation, etc. For this daring hypothesis, which at a single blow brought order and cohesion into the seeming chaos, Rutherford and his collaborators succeeded in obtaining decisive proof by isolating a great number of such transmutation products, whose chemical investigation showed that they possessed properties which corresponded exactly to the known elements, from which they only differed in that their atoms had a limited "lifetime" which sooner or later is terminated by an explosion.

In the following years, through a series of magnificently arranged and brilliantly conducted experiments, Rutherford explored the phenomena which accompany the explosion of the atoms. As an especially beautiful result of these investigations can be mentioned that it turned out that the particles emitted on the explosion had such a large kinetic energy that the action of a single emitted particle could be observed. Thus a direct method for counting atoms was obtained, whose results turned out to agree completely with those found by the methods previously mentioned. The major result achieved by Rutherford, however, was the conclusions he was able to draw about the building blocks of atoms. As the paths of the individual particles could be followed on their way through gases and other substances, it thus turned out that information could be obtained about the constitution of the atoms they met. It was thus discovered that an atom was usually very easily penetrable for the particles, as these could penetrate many thousand atoms without any noticeable change of direction. But now and again on collision with an atom the particle was found to suffer large deflection which, as Rutherford showed, must be interpreted to mean that the atom, apart from the negatively charged electrons, possessed a nucleus which was charged with positive electricity, and which contained nearly all the mass of the atom. As the nucleus is extremely small in relation to the whole atom, the latter is constructed in a way quite similar to a solar system; the atom consists of a number of light electrons which move around the heavy atomic nucleus as the planets do around the sun. With this discovery by Rutherford of the *atomic nucleus*, which was supported in many different ways, it must be assumed that all the building blocks of the atom have become known and therefore there is now a basis for penetrating into the question of the explanation of the properties of substances by means of more detailed ideas about the constitution of atoms.

While it will not be possible here to go into detail as regards the theoretical treatment of the latter question and the marvellous prospects opening for the creation of a common basis for the explanation of the phenomena of chemistry as well as physics, I will still just mention the new important result that Rutherford has reached by very recent investigations. With exceptionally well conducted and difficult experiments he has been able to observe how, on a collision between an atom and an alpha particle in which the particle directly hits the atomic nucleus – which of course happens extremely rarely – the latter can on occasion be shattered into pieces which each separately will form nuclei for atoms of other elements. Since with this result means have been found to bring forth arbitrarily a process corresponding exactly to that occurring spontaneously in the explosions of the atomic nucleus observed on the transmutations of radioactive substances, science can more or less be said to have entered a new epoch, since the stability of the atoms of elements with respect to external influence has hitherto been regarded as the solid foundation for physical and chemical considerations. The main purpose of these lines, however, has just been to try to give the reader an impression of how naturally and directly Rutherford's new results fall in with his earlier results mentioned above, through which the basic conceptions used in the chemical and physical sciences are developed into a marvellous and unsuspected clarity.

Naturally, the above remarks are only a brief sketch of some few main results of Sir Ernest Rutherford's work, which in their extent and richness of ideas are unparalleled in contemporary research in experimental physics. That it has been possible for one man alone to make such a unique achievement in science is naturally due first and foremost to his superior scientific genius, but at the same time also due to many human qualities which have enabled him to gather a great number of students around him and to guide and inspire them. It is beyond my abilities even to try to describe these qualities in detail, but this is of course not at all necessary as the Danish public interested in science will in these days have the opportunity of personally becoming acquainted with his distinctive personality. As one of the many who have had the joy of working in his laboratory under his leadership, I would just like to express the gratitude and admiration whereby Sir Ernest Rutherford is regarded by all his pupils.

N. Bohr

Niels Bohr making the broadcast for Einstein's seventieth birthday, Copenhagen, 1949.

VI. THE INTERNATIONALIST

UNESCO Courier **2** (No. 2, March 1949) 1, 7

Tribute to Albert Einstein on his 70th Birthday

THE INTERNATIONALIST (1949)

Versions published in English and Danish

English: The Internationalist
A UNESCO Courier **2,** (No. 2. March 1949) 1, 7

Danish: Niels Bohr hylder Albert Einstein
B Politiken, 15 March 1949

B is somewhat more extensive than *A*

THE INTERNATIONALIST

By
Professor Niels BOHR

IT is most natural and appropriate that the United Nations Educational, Scientific and Cultural Organization should pay a tribute to Albert Einstein on the occasion of his seventieth birthday. Indeed, for the whole of humanity Einstein's name stands pre-eminently for that search to extend our knowledge and deepen our understanding which is not only the spirit and object of science, but which forms the very foundation of all human civilization.

Through Albert Einstein's work the horizon of mankind has been immeasurably widened, at the same time as our world picture has attained a unity and harmony never dreamed of before. The background for such achievement is created by preceding generations of the world-wide community of scientists and its full consequences will only be revealed to coming generations.

Man's endeavours to orient himself in his existence beyond the immediate necessities of life may be traced back to the widely spread birth-places of our civilization like Mesopotamia, Egypt, India and China and, above all, to the small free communities in Greece, where arts and science rose to a height unsurpassed for long ages. During the Renaissance, when all aspects of human culture again flourished, most intense and fruitful contacts between scientific endeavours all over Europe took place, as we are reminded by the names of Copernicus, Tycho Brahe, Kepler, Galileo, Descartes, Pascal and Huygens, men of many countries whose achievements created the basis of the edifice of which Newton's genius is the pinnacle.

New Insight

THE great advance of natural philosophy at that time, which came to exert a deep influence on all human thinking, consisted above all in the attainment of a rational description of mechanical phenomena based on well-defined principles. It must, however, not be forgotten that the idea of absolute space and time formed an inherent part of the basis of Newton's work and that also his well-known concept of universal gravitation constitutes an element so far not further explainable.

It was just at these points that Einstein initiated quite a new development which, in an unforeseen manner, has deepened and rounded our views and given us new insight and power of comprehension.

The way leading to this turning point was paved by the development during the nineteenth century, of our knowledge of the electromagnetic phenomena which has brought such a great increase in human facilities and created the modern means of world-wide communication.

This development was furthered by an ever more active international co-operation, the extent of which is recalled by such names of many nations as Volta, Cersted, Faraday, Maxwell, Hertz, Lorentz and Michelson. Gradually, however, the growth of knowledge in this new field disclosed more and more clearly the difficulties and paradoxes inherent in absolute space-time description.

A quite new outlook was here opened by the genius of Einstein, who changed the whole approach to the problems by exploring the very foundation for the description of our experience. Thus, Einstein taught us that the concept of simultaneity of events, occurring at different places was inherently **relative**, in the sense that two such events which to one observer appear simultaneous, may seem to follow each other in time from the standpoint of another observer.

This recognition of the extent to which the account of phenomena depends essentially on the motion of the observer proved, in the hands of Einstein, a most powerful means of tracing general physical laws valid for all observers.

In the following years, Einstein even succeeded in attaining a viewpoint wide enough to embrace the gravitational phenomena, by extending his considerations to the comparison of the effects experienced by observers with accelerated movement relatively to each other. Out of Einstein's novel approach to the use of space and time concepts grew gradually a wholly new attitude towards cosmological problems, which has given most fertile inspiration for the exploration of the structure of the universe.

Although simplicity and beauty are the principal marks of Einstein's fundamental ideas, the detailed treatment of complex problems often demands the use of abstract mathematical methods like non-Euclidean geometry. As often before, it has here been most fortunate that such tools were ready as the fruit of the work of older mathematicians.

(Continued on page 7)

Internationalist

By

Professor Niels BOHR

(Continued from page 1)

The names of Gauss, Lobachevsky, Bolyai, Riemann, Ricci and Minkowski here again remind us of the fertility of international co-operation in all fields of science. And the same may be said of Einstein's other outstanding work.

For example, his explanation of the irregular motion of small bodies in liquids, based on the ideas of Maxwell, Boltzmann, Smoluchowski and Gibbs, made it possible for Jean Perrin accurately to count the atoms of which substances are built.

Dr. Bohr

We find ourselves today in a new epoch in physical science, in which experimental discoveries and theorical methods have led to a rapidity and fecundity of progress made possible only by international co-operation of an unprecedented activity and extent.

It is not possible in this occasion to disentangle the contributions of individual workers, but mention must be made of the guidance, at almost every step, which Einstein has given us by his Relativity theory and by his analysis of elementary quantum phenomena.

Altogether, this short exposition of Einstein's scientific achievements aims at giving an impression of the extent to which his originality of outlook has made him an innovator in science. At the same time, I have attempted to remind you that all scientific endeavours are parts of a great common human enterprise.

The gifts of Einstein to humanity are in no way confined to the sphere of science. Indeed, his recognition of hitherto unheeded assumptions in even our most elementary and accustomed concepts means to all people a new encouragement in tracing and combating the deeprooted prejudices and complacencies inherent in every national culture.

With his human and noble personality, characterized equally by wisdom and humour, Einstein himself has through all his life, and not least in these latter years, worked for the promotion of international understanding. On his seventieth birthday evidence of the veneration and gratitude our whole generation owes to him will reach Einstein from many sides, and we all want to express the wish that the hopes for which he has lived and worked may be fulfilled to the benefit of all mankind. .

The above photograph, taken on the day preceeding Albert Einstein's 70th birthday, shows the expounder of the theory of relativity relaxing in his home, at Princeton, New Jersey. Since 1933, Einstein has been at the Institute for Advanced Studies in Princeton.

Kirstine Meyer
f. Bjerrum

VII. KIRSTINE MEYER, B. BJERRUM
12 OCTOBER 1861 – 28 SEPTEMBER 1941

KIRSTINE MEYER, F. BJERRUM
12. OKTOBER 1861 – 28. SEPTEMBER 1941
Fysisk Tidsskrift **39** (1941) 113–115

TEXT AND TRANSLATION

KIRSTINE MEYER, B. BJERRUM
12 OCTOBER 1861 – 28 SEPTEMBER 1941 (1941)

Versions published in Danish

A Fysisk Tidsskrift **39** (1941) 113–115
B *Dr. phil. Fru Kirstine Meyer Død,* Berlingske Aftenavis, 29 September 1941

The two versions agree with each other

Kirstine Meyer, f. Bjerrum.

12. Oktober 1861 — 28. September 1941.

Paa Dr. phil. Fru Kirstine Meyers 80 Aars Fødselsdag vil alle danske Fysikere med Taknemmelighed tænke paa, hvad det for enhver af os har betydet, gennem hendes dybtgaaende historisk-fysiske Studier at blive belært om Udviklingen og Sammenhængen indenfor mange af de mest forskellige Omraader af Fysikken og om Baggrunden for de Bidrag, der her fra Landet gennem Tiderne er ydet til vor Videnskab.

Allerede hendes tidligste Undersøgelser, der blev samlet i hendes kendte Værk om Temperaturbegrebets Udvikling, giver et uforglemmeligt Indtryk af, hvilken Belæring historiske Undersøgelser kan give af de Vanskeligheder, der maa overvindes, før Begreber hentet fra Dagliglivet kan opnaa den Afklaring, som muliggør deres Anvendelse i den eksakte Videnskab. Det var ogsaa netop i Forbindelse med disse Undersøgelser, at Fru Meyer gjorde sit første Fund af Ole Rømers Optegnelser, der satte hende paa Spor efter dennes afgørende, indtil da ganske forglemte Bidrag til Fastlæggelse af Termometrenes Temperaturskala. Den Opfattelse af Rømers Indflydelse paa Fahrenheit, hvortil Fru Meyer med saa fin historisk Sans var naaet, fandt som bekendt den smukkeste Bekræftelse ved den senere Udgivelse af Fahrenheits Breve. Foruden at bidrage saa væsentligt til Klarlæggelsen af Rømers alsidige eksperimentelle Forskervirksomhed lykkedes det hende jo ogsaa at sætte selve hans store Opdagelse af Lysets Tøven i nyt Perspektiv gennem Fundet og Udredningen af de Originaloptegnelser vedrørende Jupitermaanernes Forsinkelser, hvorpaa Rømer havde baseret sine Slutninger.

Naar man tænker paa den Forskertrang og Indlevelsesevne, som Fru Meyer ved disse Undersøgelser havde lagt for Dagen, kan det ikke betragtes som nogen Tilfældighed, at det atter blev hende, der skulde fremdrage og værdsætte H. C. Ørsteds Optegnelser om de talrige og sindrige Forsøg, som han anstillede i umiddelbar Tilknytning til sin Opdagelse af Elektromagnetismen, og som førte ham til saa klar Erkendelse af mange af de elektromagnetiske Virkninger, hvis Opdagelse hidtil er blevet tilskrevet andre. Foruden at give Verden en helt ny Forstaaelse af det brede Grundlag, hvorpaa Ørsteds videnskabelige Virksomhed hvilede, har Fru Meyer som ingen anden bidraget til at belyse den gennemgribende Betydning, Ørsteds alsidige og vidt forgrenede Virksomhed har haft for Naturvidenskabens Udvikling og Udbredelse herhjemme. De varme Ord, hvormed Fru Meyer omtaler H. C. Ørsteds Opfattelse af Lærergerningen som en Kunst, der ikke alene kræver utrætteligt Arbejde med Stoffet, men som ogsaa stiller de største Krav til Selvudvikling, kan ikke undlade at faa enhver til at tænke paa den uvurderlige Indsats, som Fru Meyer netop ud fra samme Indstilling har gjort for Højnelsen af Ungdommens Undervisning i Naturfagene herhjemme.

Med disse faa Linier har det jo ikke været Hensigten at give nogen nærmere Redegørelse for de omfattende og frugtbare Studier, som Fru Meyer med utrættelig Begejstring gennem saa mange Aar har fortsat trods sin store pædagogiske Virksomhed, men blot at minde om nogle Hovedlinier i det Livsværk, med hvilket hun paa en Maade, der er uden Sidestykke, har gavnet dansk fysisk Videnskab og dens Anseelse.

Niels Bohr.

Niels Bohr Collected Works, Vols. 11 & 12

TRANSLATION

Kirstine Meyer, b. Bjerrum.

12 October 1861 – 28 September 1941.

On Dr. Phil. Kirstine Meyer's 80th birthday all Danish physicists will remember with gratitude what it has meant for each of us to be taught, through her thorough studies of the history of physics, about the development of and relationships between many of the most divergent realms of physics and about the background of the contributions which through the ages have been made from this country to our science.

Already her earliest investigations, collected in her well-known work on the development of the concept of temperature, give an unforgettable impression of the lesson that historical studies can give regarding the difficulties that must be overcome, before concepts from everyday life can achieve the clarification necessary for their application in exact science. It was indeed precisely in connection with these studies that Mrs Meyer made her first discovery of Ole Rømer's notes which put her on the track of his important, until then completely forgotten, contribution to determining the temperature scale of thermometers. This understanding of Rømer's influence on Fahrenheit, reached by Mrs Meyer with such a fine sense of history, found, as is well known, the most beautiful confirmation in the later publication of Fahrenheit's letters. Besides making such an important contribution to the elucidation of Rømer's many-sided experimental research activity, she also succeeded in placing his great discovery of the hesitation of light itself into a new perspective through the finding and the explanation of the original notes concerning the delays of Jupiter's moons on which Rømer had based his conclusions.

When we think of the desire for knowledge and ability for sympathetic insight that Mrs Meyer had shown through these studies, it cannot be considered as a coincidence that, again, it was she who should call attention to and appreciate H.C. Ørsted's notes concerning the numerous and ingenious experiments which he made in immediate connection with his discovery of electromagnetism and which led him to such a clear recognition of many of the electromagnetic effects whose discovery has hitherto been ascribed to others. Besides giving the world an entirely new understanding of the broad basis on which ørsted's scientific work rested, Mrs Meyer has contributed as no other to illuminating the decisive importance which Ørsted's many-sided and widely ramified activities have had for the development and dissemination of natural science in our country. The warm words with which Mrs Meyer refers to H.C. Ørsted's concept of the teaching profession as an art, which not only requires untiring work with the subject matter, but also places the greatest possible demands on self-development, cannot but make everyone remember the invaluable effort Mrs Meyer has made, precisely on the basis of the same outlook, for the improvement of science education for the young generation in our country.

It has not been my intention with these few lines to give a detailed account of the extensive and fruitful studies Mrs Meyer has continued with untiring enthusiasm for so many years despite her great teaching activity, but only to bring to mind some main themes in the life-work with which in a way unparalleled she has benefitted Danish physical science and its standing.

Niels Bohr.